Biological Effects
of Microwaves

The preparation and publication of this monograph was supported through the Special Foreign Currency Program of the National Library of Medicine, National Institutes of Health, Public Health Service, U.S. Department of Health, Education and Welfare, Bethesda, Maryland, under the agreement No. 05–513–1 with the Coordinating Commission for Polish-American Scientific Collaboration to the Scientific Council to the Minister of Health and Social Welfare, Government of Poland, Warsaw.

Contents

PREFACE	7
1 INTRODUCTION	11
2 PHYSICAL CHARACTERISTICS OF MICROWAVES	24
3 INTERACTION OF MICROWAVES WITH LIVING SYSTEMS	46
Penetration and Propagation of Microwaves within a Biological Target	46
Primary Interaction of Microwaves with Living Systems	63
Secondary Effects of Microwave Interaction with Living Systems	72
4 BIOLOGICAL EFFECTS OF MICROWAVES. EXPERIMENTAL DATA	78
Thermal Effects of Microwave Irradiation	78
Effects on the Nervous System	92
Cardiovascular Effects of Microwave Exposure	117
Endocrine and Metabolic Effects of Microwave Exposure	122
Influence of Microwave Irradiation on Testes, Female Genital System, Pregnancy, and Foetal Development	127
Chromosomal Effects, Possible Genetic Effects, and Influence of Microwave Radiation on Mitosis. Cellular Effects	132
Microwave Effects on Internal Organs: Chest, Abdominal Cavity, and Digestive Tract	135
Microwave Effects on Blood and the Blood-Forming System	137
Experimental Studies on Microwave Cataractogenesis	146
Miscellaneous Effects	149
Comments on Experimental Studies on the Interaction of Microwave with Living Systems	150
5 HEALTH STATUS OF PERSONNEL OCCUPATIONALLY EXPOSED TO MICROWAVES, SYMPTOMS OF MICROWAVE OVEREXPOSURE	153
6 SAFE EXPOSURE LIMITS AND PREVENTION OF HEALTH HAZARDS	170
Comments on Safe Exposure Limits	183
Prevention of Health Hazards	186
7 FINAL COMMENTS	188
REFERENCES	193
SUBJECT INDEX TO REFERENCES	228
SUBJECT INDEX	232

Contents

PREFACE

1. INTRODUCTION ... 11

2. PHYSICAL CHARACTERISTICS OF MICROWAVES ... 25

3. INTERACTION OF MICROWAVES WITH LIVING SYSTEMS ... 46
   Radiation and Propagation of Microwaves within a Biological Object ... 46
   Primary Interaction of Microwaves with Living System ... 62
   Secondary Effects of Microwave Interaction with Living System ... 72

4. BIOLOGICAL EFFECTS OF MICROWAVES: EXPERIMENTAL DATA ... 75
   Thermal Effects of Microwave Irradiation ... 76
   Effects on the Nervous System ... 82
   Cardiovascular Effects of Microwave Exposure ... 117
   Endocrine and Metabolic Effects of Microwave Exposure ... 122
   Influence of Microwave Irradiation on Hearing, Health, General System, Pregnancy, and Foetal Development ... 127
   Chromosomal Effects, Possible Genetic Effect, and Influence of Microwave Radiation on Mitosis, Cellular Effect ... 139
   Microwave Effects on Internal Organs: Chest, Abdominal Cavity, and Digestive Tract ... 153
   Microwave Effects on Blood and the Blood Forming System ... 157
   Experimental Studies on Microwave Carcinogenesis ... 166
   Ultra-Genesis Effects ... 169
   Theoretical and Experimental Studies on the Interaction of Microwaves with Living Systems ... 172

5. HEALTH STATUS OF PERSONNEL OCCUPATIONALLY EXPOSED TO MICROWAVES: SYMPTOMS OF MICROWAVE OVEREXPOSURE ... 181

6. SAFE EXPOSURE LIMITS AND PREVENTION OF HEALTH HAZARDS ... 190
   Comments on Safe Exposure Limits ... 191
   Prevention of Health Hazards ... 194

7. FINAL COMMENTS ... 198

REFERENCES ... 204

SUBJECT INDEX TO REFERENCES ... 224

SUBJECT INDEX ... 234

Preface

The present work constitutes an attempt to present to the English-speaking reader a survey of the literature on the biological effects of microwaves, including the available data on evaluation of health hazards and safe exposure limits. The authors are fully aware that the subject under consideration is highly controversial and has aroused heated discussion. It is our feeling that in the course of these discussions many misunderstandings arose from frequent misquotations, imperfect translations, and only fragmentary knowledge of the vast literature on the biological effects of microwave irradiation. Over 2000 titles could be listed as references; however, many of the papers published repeat descriptions of similar or even the same experiments. Repetitive or strictly confirmatory work was omitted from this survey. To avoid the same fault, the authors also tried to omit data already reported in the monographs by *Marha* [336] and by *Presman* [468], recently translated into English. Certain repetitions were, however, unavoidable. The reader who is especially interested in Soviet research is referred to the competent reviews by *Dodge* [127–131] and *Healer* [219]. An almost complete list of references was prepared by Z. Glaser (609) in his bibliography. The present authors have the uncomfortable feeling that at least in a few of the other existing English reviews or translations of foreign (especially Russian) work misunderstandings, arising probably from insufficient knowledge of the language and/or of the subject, may be detected. In view of this, one of us (*P. Czerski*) attempted the translation himself to prevent misunderstandings caused by insufficient knowledge of the Polish language; the authors hope that his knowledge of English will prove adequate. The native English-speaking reader is asked for forbearance.

A major cause of misunderstandings in the field of investigations on the biological effects of microwaves is insufficient multidisciplinarian collaboration. The physician or biologist usually has only limited knowledge of the intricacies of modern physics and electronic engineering. The first two chapters of this book should serve to recall certain basic data in a very simplified form and to demonstrate the need for further study. Our firm conviction is that each practical step undertaken by an industrial hygienist or physician should be checked with an engineer and a specialist in electromagnetic field propagation theory. Any biomedical research must be conducted in close collaboration with representatives of these sciences and checked with a biophysicist.

The differences in training and attitudes of mind of the interested specialists are a major source of misunderstanding in the field surveyed. An electronic engineer or physicist is able to obtain a complete quantitative description of the investigated phenomena. Such a complete description is only rarely possible in biology and medicine; many unknown or insufficiently known variables influence the investigated relationships. Statistical evaluations of the results permit the drawing of conclusions valid at a given confidence level and within the actual limits of biomedical knowledge. In certain instances it happens that a hypothesis based on questionable biological data supplemented with elegant mathematical demonstrations is accepted as scientific proof by the electronic engineer, simply because he does not realize that a given set of causally related facts was not included in the reasoning. Moreover, this set may not fit the proposed quantitative relationship. Faulty biological reasoning may lead to false conclusions in spite of the use of correct statistical methods. Conversely, the electronic engineer is apt to reject data that consist of a set of probably causally related facts on the ground that no causal relationship was proved using a quantitative description. In other words, the wide range of variation of biological phenomena and the limited possibility of giving a quantitative description of the reaction of a living system (because of its complexity) are not always appreciated by physicists and electronic engineers, perhaps with the exception of those who are interested in information theory as applied to radioelectronics.

On the other side, many biologists, and the medical profession on the whole, are insufficiently trained in mathematics and physics. The medical man is trained to react quickly to the needs of an urgent case without waiting for scientific proof of the correctness of his diagnosis. In many instances the decision must be reached quickly, and because of the short time allowed only probabilities may be considered. It is the authors' feeling that the rapid increase of electromagnetic pollution of both the working and natural environment of man should fill us with a sense of urgency.

Nevertheless, it seems that a satisfactory solution of all the questions posed by the problem of the interaction of microwaves with living systems may be obtained only by close interdisciplinarian collaboration of biologists, the medical profession, biophysists, electronic engineers, and many consultants in various specialities. The main condition of a fruitful collaboration is the realization of the limitations of each particular approach, one's own foremost.

In preparing this survey we tried to meet the requirements of the many people interested in the subject. The first two chapters are of a general introductory nature and are aimed at hygienists and physicians in industrial health service. Chapter 6, on safe exposure limits and the health surveillance of microwave workers presents the solutions proposed in various countries by the agencies responsible for the prevention of industrial health risks, with emphasis on the system adopted in our country. Certain personal views are also expressed. Chapter 4 represents a survey of the results obtained by animal experimentation and Chapter 5 the clinical findings and results of analyses of the health status of personnel professionally exposed to microwaves. Together with the list of references both these chapters may serve as a sort of annotated bibliography. In each, the final section contains an attempt at a condensed survey of the available data with a view to pointing out what may be considered as established and what questions must be posed. Such sections express the personal views of the authors and should be evaluated critically by the reader. The same concerns the short recapitulation in Chapter 7. Chapter 3, on the mechanisms of the interaction of microwaves with living systems, reports the views expressed in the literature, and the personal views of the authors on this matter, which consists mainly of doubts and question marks.

Our principal aim was to avoid any misquotations of findings or distortion of views expressed by others and to indicate clearly where personal, subjective, and probably highly controversial opinions are given.

S. Barański and P. Czerski

Chapter

1 Introduction

The term "microwave radiation" is used to designate that portion of the electromagnetic radiation spectrum which lies between the frequencies 300 and 300,000 MHz, which correspond to wavelengths in air of 1 m to 1 mm. This designation is accepted in most European countries [336, 378, 425]. Similar definitions are used in American literature [561], although the United States of America Standards Institute [468] considers the frequencies 10 and 100,000 MHz as the boundaries of the microwave region. In this book the particular spectral regions of electromagnetic radiation will be designated as shown in Fig. 1. The boundaries between successive regions were fixed arbitrarily, using the most common definitions.

A short discussion of Fig. 1, with the aim of localizing the microwave region in the electromagnetic radiation spectrum more precisely, may prove helpful. This portion of the spectrum, which encompasses the highest frequencies (the shortest wavelengths) and ends in the ultraviolet region, is called "high-energy radiation" [233] or "ionizing radiation". Because of the high photon energy of radiation from this part of the spectrum, the primary effect of its interaction with living matter consists of ionization.

Ultraviolet (UV) radiation may be considered to be the next spectral region [233]. It forms the boundary between the two large parts of the spectrum shown in Fig 1. Far UV gives biological effects similar to those induced by high-energy radiation; near UV belongs to the second part of the spectrum, "nonionizing radiation", term only now coming into wide use.

The designation of a large part of the electromagnetic radiation spectrum, comprising visible light, infrared (IR), microwaves, and radio-frequency waves by a common term, nonionizing radiation, is certainly a simplification. It seems, however to play a very useful role by indicating new tasks and perspectives in radiobiology. In the time-honored, traditional sense, the subject of radiobiology are the biological effects of ionizing radiation. The term "radioprotection" is commonly understood to pertain to the prevention of ionizing radiation health risks. The biological effects of UV, visible light, and IR are the domain of another well-established discipline, "photobiology".

Until relatively recently, the need for protection against hazards involved in the use of nonionizing radiation was negligible. The design and use of high-power sources of nonionizing radiation, which emit beams of high energy density, such as

12   Introduction

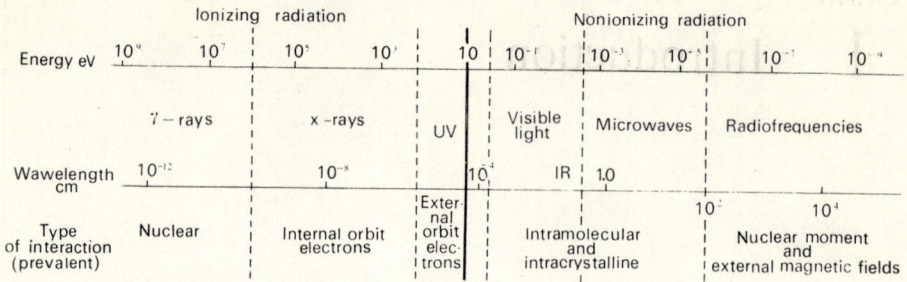

Fig. 1. Electromagnetic radiation spectrum. Boundaries arbitrary.

lasers or certain microwave apparatus, gave rise to the appearance of real hazards and the need for appropriate risk prevention. Because of the dynamic development of electronics, the number of high-intensity nonionizing radiation sources increases nearly each day.

It should be noted that many authors [336, 415, 438] designate the problems of the biological effects of microwaves and radio-frequency radiation and their health-hazard implications as problems of the biological effects of electromagnetic fields and of "protection againts microwaves" or "protection against radiowaves". The title of one of the monographs from this field, "Work Hygiene and Safety in Electromagnetic Microwave Fields", may be cited as a characteristic example [574]. Such titles reflect to a certain degree the opinions and approach of the author to the problem. In certain instances the diversity of the nomenclature may lead to misunderstandings, and one has the feeling that a part of the medical profession is disoriented and looks upon the biological effects of microwave irradiation as highly suspect and unclear phenomena, shrouded in mystery [401]. It should be stressed that the biological effects of nonionizing radiation must be evaluated in the light of modern physics, using all the possibilities offered by a biophysical interpretation of the interaction of this radiation with living systems. The starting point of all such considerations is the system

source → radiation → target

In other words, the system considered consists of a source that emits electromagnetic energy; the incident energy is absorbed and transformed within the target (irradiated object). The physical laws of optics, reflections, diffraction, dispersion, interference, quantum effects, and electromagnetic field theory are of importance and should be applied for an investigation and explanation of the observed phenomena. This statement applies of course to the whole spectrum of electromagnetic radiation.

Returning to the electromagnetic radiation spectrum as presented in Fig. 1, it may be seen that the borderline between ionizing and nonionizing radiation lies in the ultraviolet region (UV), next to which visible light and infrared radiation (IR) may

Table 1
Radio frequency and microwaves.

Approximate wavelength in air m	Frequency MHZ	DESIGNATIONS				
		Used in this work	Russian literature	Metric, used in USSR, Poland, and Czechoslovakia	English and American literature	Loosely used
$15 \cdot 10^3$ and more	0.02 and less	Very long	NTch, niskie tchastoty (low frequency)	Dekamyria and myriameter waves	VLF, very low frequency	AF, audio frequency
10^4	0.03	Long waves		Kilometer waves	LF, low frequency	
$3 \cdot 10^3$	0.1					RF, radio frequency Hertzian waves
1000	0.3	Median waves	VTch, vysokie tchastoty high frequency		MF, median frequency	
800	0.4			Hektometer waves		
100	3			Dekameter waves		
30	10	Short waves		Meter waves	HF, high frequency	
10	30	Ultra short waves	UV Tch, ultravysokie tchastoty ultrahigh frequency			
1.5	200			Decimeter waves	UHF, ultra frequency	Designated as microwaves, radar emanations, UHF, or HF
1	300					
10^{-1}	3,000	Microwaves	SVTch, sverch wysokie tchastoty superhigh frequency	Centimeter waves	SHF, super high frequency	
10^{-3}	30,000					
$1.5 \cdot 10^{-3}$	200,000			Millimeter waves	EHF, extra high frequency	
10^{-3}	300,000					
$0.75 \cdot 10^{-3}$	400,000	Intermediate and IR radiation infrared	IK, infra krasnye	IR	IR	IR
$0.3 \cdot 10^{-3}$	1,000,000					

be discerned. Microwaves constitute the next region. According to the definition adopted in this book an "intermediate radiation" between IR and microwaves (300 to 300,000 MHz) should be distinguished. Such a distinction has, however, no deeper meaning from the theoretical and practical points of view; it becomes necessary if the frequency of 100 GHz is adopted as the boundary of the microwave region. The further part of the spectrum may be divided into ultrashort waves (300 to 30 MHz, 1 to 10 m), short, median, long, and very long radiowaves (see Table 1). The microwave region may be subdivided into radar bands, as shown in Table 2. It should be noted that in this case the microwave region ends at 220 MHz, i.e., 1.333 m. Attention should be drawn to the fact that in many biomedical publications designations presented in Tables 1 and 2 may be used rather loosely.

Table 2

Microwave (radar) bands used in the United States [354] and with slight variations used internationally; supplemented by loose designations found in biomedical literature.

Wavelength in air cm	Frequency MHz	Band	Loose designation
133.3–76.9	220–390	P	1– or 1.5–m bands
76.9–19.3	390–1,550	L	50–, 30– or 20–cm band
19.3–5.77	1,550–5,200	S	20– or 10–cm bands
7.69–4.84	3,900–6,200	C	
5.77–2.75	5,200–10,900	X	3–cm band
2.75–0.834	10,900–36,000	K	
0.834–0.652	36,000–46,000	Q	Millimeter waves
0.652–0.536	46,000–56,000	V	Millimeter waves

One may easily wonder why such importance is attached to the microwave region, which constitutes only a narrow part of the whole, broad, nonionizing radiation spectrum. There are several reasons. Chronologically, microwave generators are the first nonionizing radiation sources that permitted the emission of focused beams of very high energy density. During World War II, the technique of radiolocation developed rapidly because of military requirements; this was followed by amazingly fast progress in microwave techniques. Already at this time, the first sings of concern about possible health hazards and risks to personnel occurred [106, 308]. Radiolocation equipment is constantly being perfected, and its power and the number of installations increase every day, thus leading to the present situation, where microwaves should be considered an atmospheric pollutant, similar to other industrial pollutants. In addition the introduction of color-television transmitters has caused an increase in microwave intensity in the atmosphere, since certain color TV systems work in this band. Incidentally, this new modern technique uses frequencies close to that at which waves were emitted by the first spark generator of

Biological Effects of Microwaves

STANISŁAW BARAŃSKI

Member, International Academy of Aviation and Space Medicine,
Director, Institute of Aviation Medicine, Warsaw

and

PRZEMYSŁAW CZERSKI

Member, International Academy of Aviation and Space Medicine,
Head, Department of Human Genetics, National Research
Institute of Mother and Child, Warsaw

Published by

Dowden, Hutchinson
& Ross, Inc.
Stroudsburg, Pennsylvania, U.S.A.

Copyright © 1976 by Dowden, Hutchinson & Ross, Inc. Library of Congress, Catalog Card Number: 74-7837 ISBN: 0-87933-145-3

All rights reserved. No part of this book covered by the copyrights hereon may be reproduced or transmitted in any form or by any means—graphic, or mechanical, including photocopying, recording, taping or information storage and retrieval systems—without written permission of the publisher.
Lay-out and design for printing by
Państwowy Zakład Wydawnictw Lekarskich
(PZWL—Polish Medical Publishers, Warsaw)

Printed in Poland

Benedict Hertz 455 MHz, i.e., 66 cm; however, his generator emitted waves of a decreasing amplitude (damped oscillations).

The increase in microwave pollution and possibility of health risks to personnel gave impetus to several large research projects [353]. In the 1950s and 1960s many investigations on the biological effects of microwaves were carried out in numerous research centers in several countries, among them France, Poland, Czechoslovakia, the United States and the USSR. During the last 30 years, a vast amount of experimental data and clinical observations was collected, and an attempt at a survey of this material seems timely.

During the last 10 years industrial applications of microwave techniques expanded greatly, and microwave heating entered into common use. Thus microwave techniques invaded private homes, where more and more microwave ovens were installed. This was a cause for public anxiety, alarm even, and led to polemics full of controversial statements. Discussions may be found in Electronics during the years 1969-1971 (see, e.g., [402, 568]). The general interest in the biological effects and health implications of microwaves induced us to attempt this survey of the relevant literature.

Table 3 presents selected examples of the most common uses of microwave techniques and equipment. It should be added that many nonionizing radiation sources widely used in industry, radiolocation, radiocommunication, and medicine (shortwave diathermy) emit waves of lower frequencies, 300 to 30 MHz. Many statements and findings concerning the biological effects of microwaves may be applied also to waves of this region, which is typical for television, one of the most common media of mass information. The increasing use of industrial equipment that generates waves of frequencies below 30 MHz should also not be forgotten. In actuality, frequencies of a few megahertz seem to be most important. Nevertheless, the increasing industrialization of the world and the tendency to increase the power of equipment, a veritable power race, incline one to consider the attempts at limiting microwave exposure as a possible model for the solution of the problem of nonionizing radiation hazards in the whole radio-frequency range, including low and very low frequencies. Because of this, Table 3 was supplemented by examples of the use of sources emitting radiations in frequency ranges other than the microwave. More data on the use of such equipment and related biomedical problems may be found in monographs by *Marha* [336] and *Presman* [438]. Textbooks on medical diathermy may also be consulted [472, 477, 565]; see also "reviews" in the Subject Index to References.

A detailed presentation of the application of microwave techniques and a description of the equipment lie outside the scope of this work. A few very simplified and brief remarks, based on *Panecki* [417], will be given.

One of the oldest, but still in wide use, microwave transmitter valves is the magnetron (Fig. 2). It contains two coaxially placed electrodes, thus ensuring circular symmetry. As in other electronic tubes, the cathode is heated by an electric heater,

Table 3.

Selected examples of typical uses of equipment generating radio-frequency and microwave radiation (based on [80, 354, 378]).

Frequency	Use	Occupational exposure	Examples of potential incidental exposure — general population hazards
Below 3 MHz	Metallurgy: welding, melting, tempering, etc. Broadcasting, radiocommunications, radionavigation	Various factory workers, e.g. furniture veneering operators, drug and food sterilizers, auto industry workers	Factory executive personnel, watchmen, guards
3–30 MHz	Many industries e.g., auto, wood, chemical, food. Heating, drying, welding, gluing, polimerization, sterilization of dielectrics. Agriculture, food processing, medicine, radioastronomy, broadcasting	Electronic engineers and technicians: air crewmen, missile launchers, radar mechanics and operators, microwave-oven operators and maintenance workers	Airport and seaport personnel of various professions; inhabitants of areas in the vicinity of high-power radar installations, broadcasting stations, and TV transmitters
30–300 MHz	Many industries as above. Medicine. Broadcasting. TV, air traffic control, radar, radionavigation	Scientists, physicists, microwave development workers	
300–3,000 MHz	TV, radar (tropooscattor and meteorological). Microwave point-to-point. Telecommunication, telemetry. Medicine. Microwave ovens. Food industry	Microwave testers; diathermy, microwave diathermy, operators and maintenance workers: medical personnel	Housewives and children (microwave ovens in private homes)
3–30 GHz	Altimeters, air- and ship-borne radar, navigation, satellite communication, microwave point-to-point	Broadcasting transmitter and TV personnel	
30–300 GHz	Radioastronomy, radiometeorology, space research, nuclear physics and technique, radio spectroscopy	Marine and coastguard personnel, sailors, fishermen, persons professionally present on board ships	

which leads to the emission of free electrons into the space between the electrodes. A constant magnetic field, perpendicular to the trajectory of the electrons, is generated by a magnet or an electromagnet. This field causes changes in the trajectory of the electrons before they are eliminated through the anode (electric current flow), and electromagnetic waves are generated. A metal loop (coupling loop)

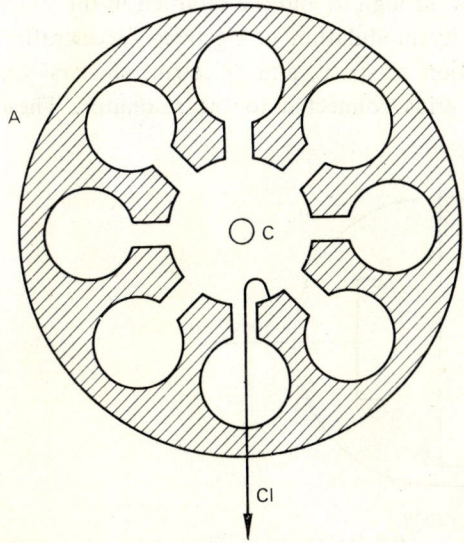

Fig. 2. Magnetron, schematic representation. A, anode; C, cathode; Cl, coupling loop. See the text. From [417], by permission.

introduced into one of the cavities permits the transmission of the field to external circuits. The dimensions of the cavities, cathode, and coupling loop of the magnetron as well as the voltage applied to the electrodes and the intensity of the magnetic field, may vary according to the desired frequency and power output. The cavities in the metal block of the magnetron are equivalent to a resonant circuit consisting of a certain inductance and capacitance. The smaller the dimensions of the magnetron and its cavities, the higher is the frequency of the generated ascillations. To obtain continuous wave (CW) generation, a constant value of the applied voltage is used. If it is switched on and off for very brief periods of time (pulses), the oscillations within the magnetron become pulsed (pulse modulation). It should be kept in mind that the power output during such a pulse (peak power) is many times higher than the mean power output. The difference may attain several orders of magnitude.

Besides microwave, one important fact to remember is that ionizing radiation (X-rays) is also generated. Whenever high-energy electrons collide with matter, X-rays are generated. The energy of these rays depends on the voltage and the intensity of the current (amperage). Depending on the mode of operation, continuous or pulsed X-rays are generated. Incidental X-ray generation occurs also in other microwave tubes.

Another type of microwave tube is the klystron. Two kinds may be mentioned: low-power reflex klystrons and multicavity (multiresonator) klystrons of high power output, up to tens of megawatts. Klystrons operate on the principle of electron velocity modulation. A reflex klystron is shown schematically in Fig. 3. A positive potential is applied to the grid of the resonator cavity RC. The heater H causes electron emission from the cathode. Because of the distribution of potentials, the electrons are directed toward the grid, pass through it, and are retarded in the space between the grid and the reflector. Velocity modulation of electrons (acceleration and retardation) leads to their redistribution in space, and "electron clusters" are formed. These are finally eliminated by grids connected to the resonator. These

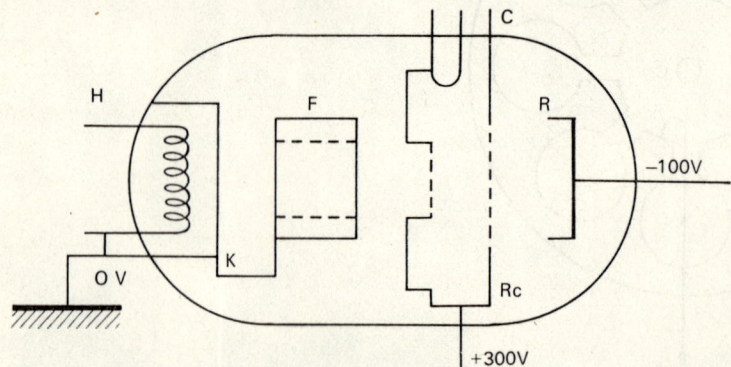

Fig. 3. Reflex klystron, schematic representation. Rc, resonator cavity grid; R, reflector; H, heater; K, cathode; C, coaxial line with a coupling loop next to it. See the text. From [417], by permission.

phenomena cause oscillations within the resonator, which is the equivalent of a resonant circuit consisting of a certain inductance and capacitance. The dimensions of the resonator and the voltage applied to the electrodes determine the frequency of oscillations, which are transmitted through a loop and the coaxial line C.

A multicavity klystron is represented schematically in Fig. 4. Two resonator cavities are shown but their number may vary. The heated cathode C emits free electrons, which are accelerated or retarded during their passage through the first

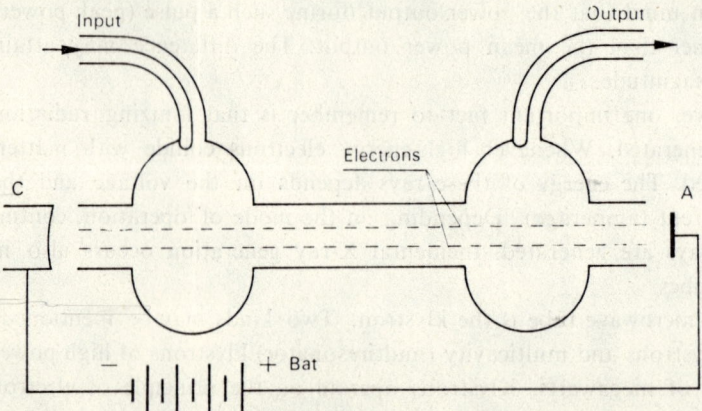

Fig. 4. Multicavity klystron, schematic representation. A, anode; C, cathode; Bat, source of current. See the text. From [417], by permission.

cavity by a high-frequency signal from an external generator. On the path between the resonator cavities, clusters of electrons are formed. As they pass through the next cavity, they generate electromagnetic waves of the same frequency, but of a higher amplitude. The oscillations are transmitted from the output resonator by a loop and a coaxial line. The electron stream is focused axially by a constant electric or magnetic field generated by a magnet or an electromagnet.

Traveling-wave tubes are also used for the generation or amplification of microwaves. The principle of operation is similar to klystrons, i.e., interaction between an electron stream and electromagnetic waves along a certain path. The electrons in the stream and the electromagnetic wave may have the same direction of propagation (forward wave tube) or opposite directions (backward-wave tube). Figure 5 is a simplified schematic drawing of a traveling-wave tube. A direct-current circuit may be seen: electron ejector E, collector Col, and a direct-current source Bat. The second circuit, a high-frequency circuit, consists also of several elements: the input circuit 1, the spiral retarding line Sp (helix), and an output circuit a coaxial line 2, the internal cable of which is connected to the helix within the tube. The retarding helix permits the adaptation of the electron stream velocity and propagation velocity of the electromagnetic wave. On this depends the possibility of interaction between the electron stream and the electromagnetic fields, i.e., generation and amplification.

To increase the stability of the tube, attenuating material Att is added. The collector Col corresponds to the anode, eliminating the electrons emitted from the ejector. The magnet, which generates the constant magnetic field that focuses the electron stream, is not shown in Fig. 5. Many types of traveling-wave tubes exist, among which are the widely used backward-wave tubes called "carcinotrons".

Microwave tubes and valves are constantly being perfected. Various semiconductor and solid-state elements are widely used. Novelties of 2 years ago seem hopelessly

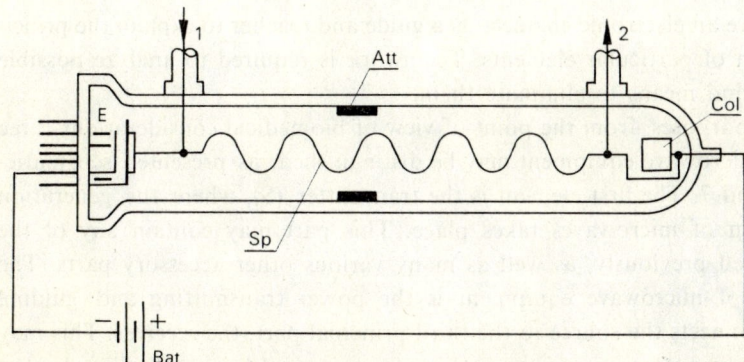

Fig. 5. Traveling wave tube, schematic representation. Sp, spiral retarding line; Col, collector; E, electron ejector; Bat, direct-current source; 1, input circuit; 2, output circuit connected to a coaxial line; Att, attenuating material. See the text. From [417], by permission.

old-fashioned now and the above-described elements, used universally in the early 1960s seem almost venerable in their "old age" in 1976. In the present authors' opinion, the practical industrial physician or hygienist must have at least a general

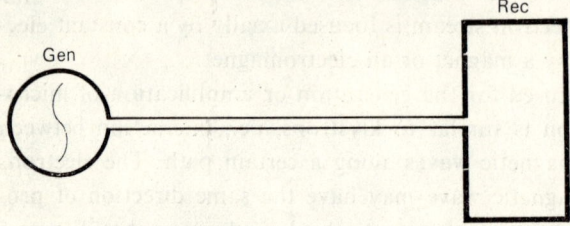

Fig. 6. Microwave system, schematic representation. Gen, transmitter, where generation and amplification take place; Rec, receiver; both these parts are connected by a transmission line. From [417], by permission.

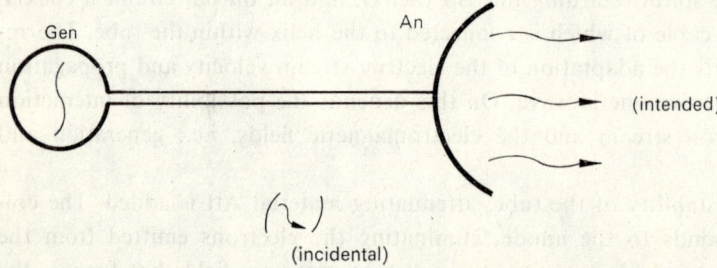

Fig. 7. Schematic representation similar to Fig. 6. A radiating element has been substituted for the receiver, the antenna An. This is emitting intended radiation into the surrounding environment. Incidental (nonintended) radiation may be emitted by the transmission line power leakage. From [417], by permission.

idea about the equipment used by the personnel entrusted to his care. The best solution is to have an electronic engineer as a guide and teacher to explain the principles of operation of particular elements. Teamwork is required to analyze possible hazards and to find means to eliminate them.

For practical purposes, from the point of view of biomedical considerations three basic parts of microwave equipment may be distinguished, as presented schematically in Figs. 6 and 7. The first element is the transmitter (S), where the generation and amplification of microwaves takes place. This part may contain any of the elements described previously, as well as many various other accessory parts. The second element of microwave equipment is the power transmitting and guiding system, which connects the source to the third principal part, the receiver. This may absorb all the transmitted energy, as in Fig. 6, or it may be a radiating element (antenna) that emits radiant energy into its surroundings (Fig. 7).

For hygienic considerations in the use of microwave equipment, an arbitrary distinction of intended and nonintended radiation may be introduced [573]. Such distinction has proved to be helpful in our practice and will be used here when examining the relationship between the microwave radiation source and the irradiated biological object (target). The radiation of the frequency, direction of propagation, and point of origin that conform to and are determined by the destination (intended use) of the equipment will be called intended radiation. Because of physical laws and in spite of the perfection of the design of the transmitter, nonintended radiation may be generated. This radiation is not put to any use; from the technical point of view it only represents energy loss. The frequency, direction of propagation, and point of origin are variable and may vary in time. From the biomedical point of view, especially in the hygienic consideration of the prevention of hazards to personnel, nonintended radiation constitutes one of the principal sources of health risks. Ionizing radiation generated incidentally in the transmitter in microwave generating tubes, amplifiers, and rectifiers is a typical example of what is called here nonintended radiation. Both intended and nonintended radiation are predictable, the latter, however, only to a certain, limited extent.

In the system presented in Fig. 6, no intended radiation is emitted outside the system by definition; the receiver is designed to absorb all the energy generated within the transmitting part. Ideally, microwave ovens should be such closed systems with no nonintended radiation present outside the equipment.

In the alternative system presented in Fig. 7 the radiating element sends out intended radiation; i.e., electromagnetic waves are emitted outside the equipment to a purpose and knowingly. Typical examples of such sources are radar installations and telecommunication or television transmitters that operate in the microwave region.

In the Figure 6 the power distributing and guiding system is not a source of radiation emitted outside the equipment. Ideally, no radiation should be detectable in the proximity of the guiding elements. In practice, such elements are a constant source of various radiations, and the problem of leakage is one of the most important questions in the prevention of health risks to personnel, when the principal aim is to limit radiation exposure to the minimal unavoidable level. The experience of the present authors demonstrates that nonintended radiation is insufficiently considered or even completely omitted in the evaluation of exposure or the analysis of health risks at workplaces. In certain instances nonintended radiation was not taken into account when experimental situations for research purposes were devised or when the bioeffects of microwaves were evaluated on the basis of animal experimentation.

Before microwave power distributing and radiating elements are considered, a few data on the physical characteristics of microwaves must be recalled; Chapter 2 is devoted to these problems. We conclude this introductory chapter with a discussion of Fig. 8, which emphasizes the main subject of this survey: the interaction between incident radiant energy and irradiated biological systems, i.e., the biological effects of microwaves and the eventual health risks. The source of the microwaves lies outside our interest; appropriate data may be found in physics and engineering textbooks. The physical characteristics of the radiation, its frequency, propagation in space, energy density, etc., are important factors. Nevertheless, the main object of this survey is to present the biological effects, i.e., the results of the interaction of microwaves with living systems. The final aim of the research conducted

Fig. 8. Schematic representation of the subject of this survey, the relationship between the source of microwaves, the radiation generated by it, and the biological target. Source characteristics, propagation of the radiation in the environment, penetration of radiation into the biological object, and energy absorption inside it must all be taken into account in considerations of the biological effects of microwaves.

in this field is to understand the mechanisms underlying the observed effects. It should be kept in mind that at present this is not always possible. A great body of phenomenological observations exists. These observations cannot be omitted and rejected because the underlying mechanisms remain obscure and at present are unexplained. Doubts and objections should serve as a basis for further research aimed at clarification and quantitation of the suspected relationships.

One final point should be made. In discussing the biological effects of microwaves, several physical terms are often misused. Most authors, radiation protection guides, and official orders use the term "power density" or "energy density". In standard internationally accepted nomenclature (Système International, SI), this refers to power or energy per unit volume. In common usage, however, power or energy density is the energy or power density per unit area in the case of microwave irradiation of living systems. In discussing the relationship between radiation and the biological object, we use the generally accepted, although incorrect, exposure dose expressed as power density per unit area, which corresponds to energy density flux (radiant energy in units of joules) per unit area. It should be stressed that analogies from ionizing-radiation research or photobiology (photometric terms) are inapplicable in the case of the biologic effects of microwaves. It remains for future research to give a more precise physical meaning to what a biologist or physician would term the relationship of exposure to absorbed dose.

Chapter 2 | Physical Characteristics of Microwaves

The aim of this chapter is to recall in a brief and simplified form the basic physical characteristics of microwaves, following the example of similar literature surveys. The biologist, the physician, or the industrial hygienist pressed by professional duties may easily forget his basic physics learned during early training. Because of this, certain basic notions are presented at the beginning of this chapter. Readers better versed in this subject are advised to skip this part.

Some readers may find this presentation useful in the evaluation of the text of later chapters or even helpful in the course of practical activity in the field of prevention of microwave hazards. If such is the case, the "biomedical" reader is encouraged to verify his knowledge and broaden it and to use a physicist or electronics engineer as a consultant in selecting additional reading matter, according to individual needs. The present authors used several physics and engineering texts, among which a few are readily accessible to the English-speaking reader [4, 97, 287, 293, 408].

As mentioned before, the terms "very high frequency electromagnetic fields" and "microwaves" are used alternatively in the literature. The reciprocal relationship may be expressed as follows [566]:

1. Changes within the components of the electromagnetic field propagating through space constitute electromagnetic waves.

2. The electromagnetic field occurs at a given time at the point in space (area) where the propagating wave actually passes.

The electromagnetic field consists of interrelated electric and magnetic forces (alternating electric and magnetic fields). The electric component is determined by the potential difference per unit length, and the magnetic component by the distribution of magnetic forces per unit length. Both components are vectors, i.e., characterized by a determined direction.

The electric field strength or electric field intensity E is defined as the force per unit charge exerted upon a test charge placed at a point; it represents the value of the vector of the electric field and is measured in volts per meter (V/m). The displacement of a positive test charge (proton) indicates the direction of the vector. Electric field force lines are indicated by the paths of charges. Similarly, the force exerted on a unit magnetic pole is the measure of the magnetic field intensity H in amperes per meter (A/m). The field force lines are determined by the orientation of a unit magnetic pole or by the direction taken by a unit current.

Each change in the electric field is accompanied by a change in the magnetic field. The converse is also true. The propagation of such changes constitutes the electromagnetic wave, or, in other words, the electromagnetic wave consists of oscillations of the vectors E and H propagating through space i.e., of alternating electric and magnetic fields [287]. The truth of these statements was demonstrated by Maxwell's equations.

Two basic phenomena underlie the formation of an electromagnetic wave:
1. Electromagnetic induction.
2. The appearance of a magnetic field (magnetic effects) around a conductor carrying electric current (moving charges).

The movement of a magnet induces an electric current in a coil made from insulated wire (induced current). Depending on the direction of movement of the magnet relative to the coil, toward or away from it, the direction of the current flow changes. Thus when the wire (conductor) is placed in an alternating magnetic field, an electric current, i.e., an electron stream moving along the conductor, is induced. Movement of electrons occurs when the conductor is placed within an electric field. It may be reasoned that the alternating magnetic field induces an electric current, because it induces an electric field. Conversely, an electric current induces a magnetic field. In the case of an electric current flowing along a straight wire, the magnetic field force lines form coaxial circles and, in the case of a coil, overlie one another (Fig. 9). The direction of the magnetic field vector is determined by what is called the right-hand-screw rule or the corkscrew rule: if a right-hand screw is driven in the direction of the current flow, the direction of the rotary movement of the screw corresponds to the direction of the magnetic field forces.

The preceding examples may be generalized. Figure 10 shows electric and magnetic field force lines between two parallel capacitor plates. The induced field is not limited to the space in which the inducing field exists, but it spreads out into surrounding

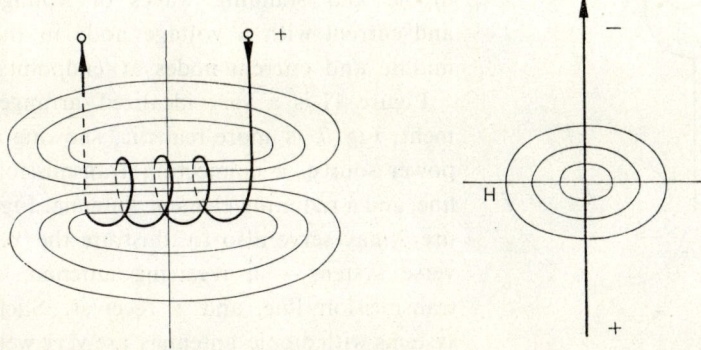

Fig. 9. Magnetic field force lines around a coil or straight conductor through which an electric current flows, the direction being indicated by the arrowhead i.e., from + to —. From [556], by permission.

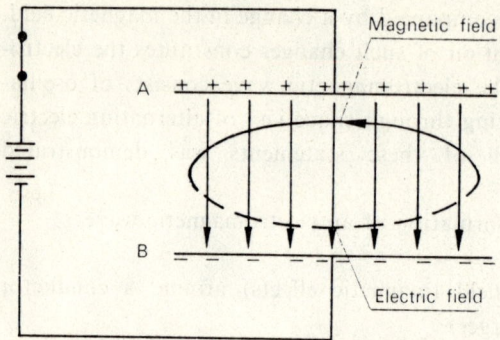

Fig. 10. Electric and magnetic field force lines between capacitor plates. From [556], by permission.

space. In this manner the alternating "conversion" of the fields occurs farther out from the source, and an electromagnetic wave is generated and sent out. Taking Fig. 10 as a starting point and remembering that a magnetic field is formed around a conductor carrying an electric current, the simplest type of transmitter may be obtained (Fig. 11). This transmitter consists of an electric current generator coupled with two straight wires (a dipole antenna). The alternating electric field generated by the transmitter is the source of electromagnetic wave generation. Such a transmitter as depicted in Fig. 11 may be referred to as an open source of oscillations. Half-wave dipole antennas in the form of a conductor one half-wave long and connected at its middle to a two-wire transmission line are commonly used. The electron stream moves rhythmically in the dipole and standing waves of voltage and current with a voltage node in the middle and current nodes at endpoints.

Figure 11 is a very idealized arrangement; Fig. 7 is more realistic, showing a power source, a connecting transmission line, and a radiating element antenna. Figure 7 may serve also to illustrate the reverse system — a receiving antenna, a transmission line, and a receiver. Such systems with dipole antennas are very well known from everyday experience; it suffices to visualize for a moment rooftop television antennas.

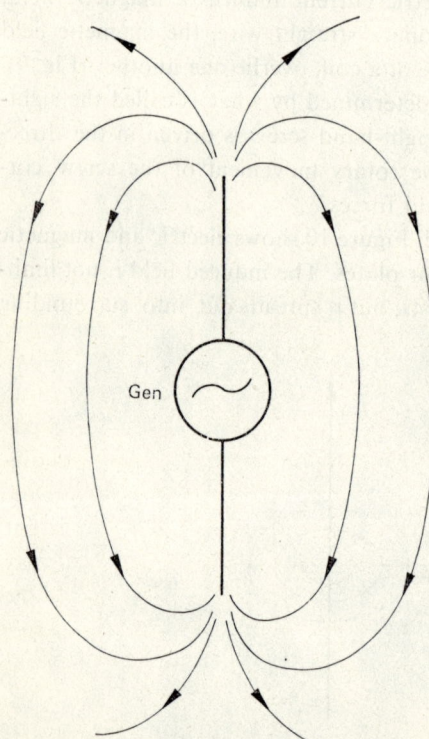

Fig. 11. Schematic drawing of a "Basic" transmitter obtained by transposition of Fig. 10. See the text. From [556], by permission.

The preceding description is extremely simplified; knowledge of other physical facts and laws, as well as a certain mathematical apparatus, is necessary to obtain a more meaningful picture of electromagnetic wave generation and emission. The simplified data on the following pages constitute the absolute minimum for a starting point for additional study.

Electromagnetic waves may be characterized by a set of parameters. One of these is frequency or its reciprocal, the period:

[1] $$f = \frac{1}{T}$$

where f = frequency, expressed usually in cycles per second or hertz 1 Hz = 1 s^{-1}; in the microwave range it is more convenient to use multiples of this unit: megahertz (1 MHz = 10^6 Hz) or gigahertz (1 GHz = 10^9 Hz),
T = period expressed in time units.

The electromagnetic wave has an amplitude, which may be characterized by the electric field intensity E or by the intensity of the magnetic field H. The time relationship of the oscillations is the phase φ, which may be expressed in degrees or in multiples of π. Angular frequency may also be used. The electromagnetic wave travels (propagates) with a given velocity and has a determined length. The following relationships exist:

[2] $$f = \frac{v}{\lambda}, \quad \lambda = \frac{v}{f}, \quad \omega = 2\pi f$$

where f = frequency,
v = velocity,
λ = wavelength,
ω = angular frequency.

The velocity of propagation of the electromagnetic wave through a given medium depends on its electric and magnetic properties, which may be determined using several parameters. One is the dielectric constant (coefficient) ε'. This is a dimensionless parameter and represents the ratio between the capacitance of a capacitor, between the plates of which the considered dielectric was introduced, and the capacitance of the same capacitor in vacuum (in practice, also in air). The absolute dielectric constant ε_0 (permittivity of free space or specific inductive capacity) is a universal constant and has the value 8.85 picofarads per meter (pF/m). Another important parameter is permeability, which indicates the ease with which a magnetic field may be set up in the medium. Relative permeability μ' is also dimensionless and is the ratio of the permeability of the medium to that of free space. Permeability of free space μ_0 has the value of 1.257 · 10^{-6} henry per meter (H/m). The velocity of the electromagnetic wave in a given medium may be expressed as

[3] $$v = \frac{c}{\sqrt{\varepsilon' \mu'}}$$

where c is the velocity of light in free space, approximately 300,000 km/s.

As the relative dielectric constant and relative permeability of air may be considered as having the value 1.0, the relationship between frequency and wavelength in air may be simplified and the following expression used:

$$\lambda = \frac{300}{f}$$

if λ is measured in meters and f in megahertz.

Biological substances (body tissues) have various dielectric and magnetic parameters. The wavelength in biological media is different from that in air, as shown in Table 4. It should be mentioned here that the dielectric constant of the medium

Table 4

Electromagnetic wavelength (cm) in air and in animal tissues at various frequencies (MHz); according to [160], slightly modified.

Medium	Frequency					
	400	1000	3000	10,000	24,000	35,000
Air	75	30	10	3.33	1.67	0.86
Skin	10.12	4.41	1.49	0.51	0.25	
Fat	30.9	12.42	3.79	1.45	0.68	
Muscles	9.41	3.09		0.62		
Blood	8.89	3.87	1.36	0.45	0.21	0.17
Bone marrow	32.19	12.63	3.97	1.25	0.37	0.39
Nervous tissue brain	11.16	4.97	1.74	0.59	0.20	0.20
Lens	12.53	5.28	1.75	0.57	0.20	0.20
Vitreous body	7.96	3.41	1.18	0.39	0.15	0.15

depends on the frequency of the external field, so the calculation of wavelength in biological media is somewhat complicated (see Chapter 3). Moreover, the expressions given here concern only lossless media. The properties of a lossy material are specified by the complex dielectric constant ε^* and complex permeability μ^*, consisting of real and imaginary parts, according to the expressions

[4] $$\varepsilon^* = \varepsilon' - j\varepsilon''$$

[5] $$\mu^* = \mu' - j\mu'' = \mu_0$$

where ε'' and μ'' are loss factors and j is $\sqrt{-1}$ imaginary unit.

Since biological material are nonmetallic, μ^* is equal to μ_0.* In this book the complex dielectric constant will be designed ε, its real part, ε' remembering that

---

* This is a simplification; it may be assumed that biological materials are diamagnetic, although paramagnetic or even ferromagnetic biologically active molecules may be encountered in living systems.

for lossless media both are equal, the loss factor, ε'', and the permeability, μ equal to μ^*, μ', and μ_0 for the materials considered.

Another property of the medium influencing the electromagnetic wave is conductivity, which may be expressed as the reciprocal of ohms per meter (Siemens per meter S/m). The relationship between the wavelength in free space λ_0 (in practice, in air) and the wavelength in the medium λ_m is defined by the expression

$$[6] \qquad \lambda_0 = \lambda_m \left[\frac{\mu}{2} \sqrt{\varepsilon'^2 + \left(\frac{4\pi\sigma}{\omega}\right)} + \varepsilon' \right]^{1/2}$$

where ε', μ and σ (conductivity) specify the medium, and ω is the angular frequency of the wave.

Changes in the interrelated characteristics of the electromagnetic wave (length, frequency, velocity) on its passage through the medium are basic for consideration of the propagation and absorption of the energy of microwaves in a living system. This is usually a multilayered object with complex geometry, consisting of various tissues with different electric properties. In consequence, these phenomena are important for evaluating the biological effects of microwaves (see Chapter 3).

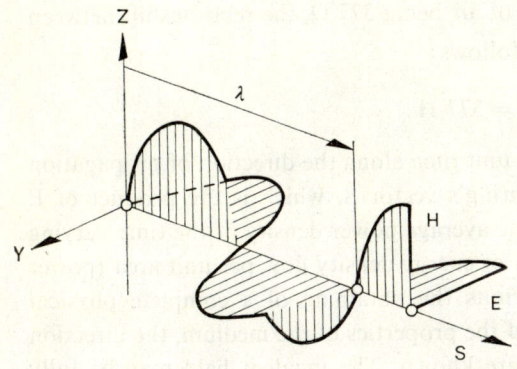

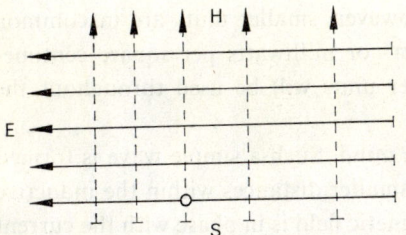

Fig 12. Propagating simple plane wave, schematic representation. H, magnetic component vector; E, electric component vector; S, arrow indicating the direction of propagation (Poynting's vector) along which power density measurements are made; λ, one wave lenght. From [336] and [293], modified slightly.

The electromagnetic wave propagating in air may be considered in its simplest form. It may be assumed that the propagation takes place in free space, i.e., in an unbounded isotropic, homogeneous, linear medium.* The electric and magnetic

---

* The dielectric constant does not depend upon the direction of flux density (isotropic), upon its magnitude (linear), and upon the location of the flux density (homogenous); no polarized particles are present.

field vectors are perpendicular to each other and to the direction of propagation according to the right-hand-screw rule mentioned earlier. Such a wave, represented in Fig. 12, may be termed a simple plane wave. $\overline{E}$ and $\overline{H}$ are dependent on the intrinsic impedance of the medium (the ratio of voltage to current traveling together in a particular direction). This is defined by the square root of the ratio of the complex permeability to the complex dielectric constant:

[7] $$Z = \frac{|\overline{E}|}{|\overline{H}|} = \sqrt{\frac{\mu}{\varepsilon}}$$

or by the expression

[8] $$Z = \frac{|\overline{E}|}{|\overline{H}|} = \left(\frac{\mu}{\varepsilon} - j\frac{\sigma}{\omega}\right)^{-\frac{1}{2}}$$

which shows the dependence of impedance on permeability, dielectric constant, and conductivity of the medium; ω is the angular frequency and j the imaginary unit equal to $\sqrt{-1}$. The intrinsic impedance of air being 377 Ω, the relationship between the vectors $\overline{E}$ and $\overline{H}$ may expressed as follows:

[9] $$\overline{E} = 377\,\overline{H}$$

The radiated energy per unit area and unit time along the direction of propagation may be expressed as the complex Poynting's vector S, which is the product of $\overline{E}$ and $\overline{H}$. The vector S is related to the time average power density of the time varying electromagnetic wave. The measurement of energy density flow per unit area (power density) in these simple conditions permits the obtaining of a complete physical definition of the conditions of exposure if the properties of the medium, the direction of propagation, and wave polarization are known. The incident field may be fully characterized and values of E and H computed. The measured power density may be expressed in watts per square meter W/m². However, smaller units are in common use: microwatts per square centimeter μW/cm² or milliwatts per square centimeter mW/cm². To avoid confusion, these smaller units will be used throughout the text.

A very important restriction must be kept in mind. Such a simple wave is formed only at a certain distance from the source. At smaller distances within the inductive field, E and H may be shifted in phase. The magnetic field is in phase with the current in the antenna and the electric field with the electric charges on it. An energy flow back and forth between these fields may occur. The electric and magnetic components do not show simple proportionality and should be measured separately. In view of this, a far-field zone and a near-field zone should be distinguished. More complicating phenomena may occur. Multipath interference, reactive near-field components, interaction between the source and nearby objects, complicated modulation of the field, polarization, and, in the case of irradiation of biological objects, subject-source coupling, make the usual dosimetry of microwave exposure questionable in many instances. For a more detailed and far more sophisticated discussion of these problems, see the papers by *Wacker* and *Bowman* [49, 50, 563, 564]. Here it suffices to say that the universally accepted quantification of microwave exposure (measurement of incident radiation in relation to the observed bioeffects) by measurement of the mean power density is not valid for the near-field zone and is somewhat of a simplification in many other instances.

Such plane-wave approximations as just described may be used for practical purposes at distances from the source that are large compared to the ratio of the squared greatest distance between the points in the source (reflector diameter in reflector antennas) to the wavelength. The extent of the near-field zone is often computed from the expression

[*10*] $$D_{N/F} = \frac{2a^2}{\lambda}$$

where $D_{N/F}$ = distance separating the near- and far-field zones,
a = greatest distance between points in the source.

This expression is valid only for uniformly illuminated, constant-phase aperture sources in the direction of maximum radiation, a point stressed by *Wacker* and *Bowman* [564]. The practical conclusion for the biologist or the medical man is to secure competent advice before accepting a value as a measurement of microwave exposure in a particular instance.

Another characteristic of the electromagnetic wave is its polarization. This is determined by the electric field vector, which is perpendicular to the direction of propagation. A plane drawn through the vector of electric field intensity and the direction of propagation is referred to as the plane of polarization. If the vector E remains in a constant plane drawn through the direction of propagation, the polarization of the wave is referred to as linear. Antennas of special design allow waves with the polarization plane rotating around the direction of propagation. If the value of the vector E is constant, the polarization is circular; if it varies periodically, the polarization in elliptical. The latter is used often in radiolocation. In other words, linear polarization may be horizontal (the electric field vector is in the horizontal plane) or vertical (the electric field vector is in the vertical plane). If equal-amplitude horizontal and vertical polarized waves 90 degrees out of phase are combined, the wave has circular polarization; if the amplitudes are not equal, the wave has elliptical polarization.

The basic points on electromagnetic waves are best summarized according to *Lance* [293] as follows:

1. Time-varying currents are the source of electromagnetic waves.
2. The electromagnetic wave may be represented as a vector field consisting of two components, the electric vector and magnetic vector.

3. Propagation of fields from the source is radial in wave fashion; a change in the media of propagation affects this phenomenon according to the dielectric constant, inductivity, and permeability of the medium, from the biological point of view permeability may be practically neglected in biological media.

4. Partial transmission and partial reflection of the electromagnetic wave occur at the boundary of media with different electric properties.

5. Electric field lines may begin and end on charges; if an electric field ends on a conductor, it must represent a charge induced on conductor.

6. Magnetic field lines cannot end; as no magnetic charges are known physically, magnetic fields form continuous closed loops surrounding a conduction current or a changing electric field.

7. Lines indicating the direction of propagation may be referred to as rays.

Using the power density concept and considering microwaves as rays, one is apt to think in terms of geometrical optics. All the phenomena of diffraction, interference, reflection, dispersion, and absorption of microwaves in various media exist. Geometrical optics must be replaced, however, by physical optics, because microwaves are much longer than visible light rays and so much longer in relation to the irradiated objects. This is illustrated by Fig. 13, where the real shadow of an object that will not pass microwaves (a screen) casts a much smaller real shadow, than the geometrical one because of the diffraction of relatively long waves in relation to the object. This sort of optics may be applied only in the far-field zone; in the near-field zone it may be inapplicable. Moreover, in certain instances such reasoning must be replaced by considerations based on electromagnetic field theory. For example as presented by *Wacker* and *Bowman* [564], a standing wave formed by two single plane waves of equal amplitude, same polarization, and opposite directions has an average-time zero power density, but the electric and magnetic energy densities may be quadruple those of the original waves at certain locations. In the near-field zone the average-time power density does not represent the electric and magnetic field intensities.

When considering electromagnetic rays, the quantum theory should also be remembered. An elementary photon of the radiation has a quantum energy E depending on frequency, according to the following expression:

$$E = hf$$

where h = Planck's constant equal to $6.6 \cdot 10^{-34}$ Js.

The magnitude of the quantum energies of electromagnetic radiations is expressed in electronovolts (eV). The quantum (photon) energy of radiation increases with frequency, as shown in Table 5. The magnitudes presented illustrate the basic difference between ionizing and nonionizing radiation, UV being the boundary region. The quantum energy of the microwave photon is not sufficient to cause ionization in the irradiated living system. This important property must be considered to avoid

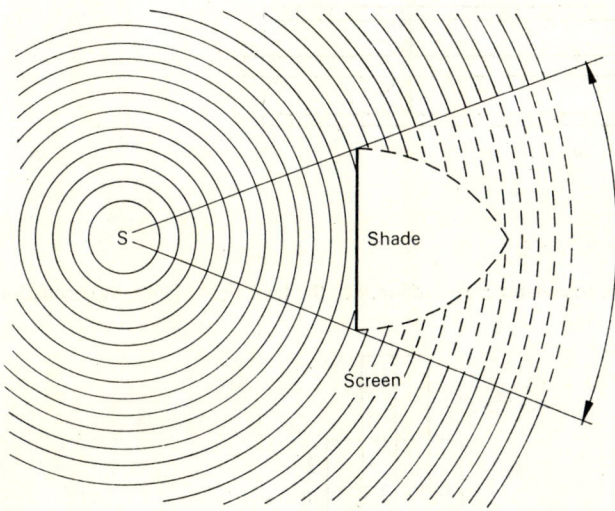

Fig. 13. Schematic representation of electromagnetic wave diffracttion effect on the dimensions of the shade cast by a nontransparent screen. S—source. From [574], by permission.

Table 5

Photon energy (eV) and frequency (MHz) of selected bands of the electromagnetic radiation spectrum; according to [21], slightly modified.

Photon energy order of magnitude	Frequency	Designation
10^5–10^{11}	$75 \cdot 10^{13}$–$3 \cdot 10^{15}$	Ionizing radiation gamma
10–10^5	$6 \cdot 10^{10}$–$75 \cdot 10^{13}$	X-rays
10–10^2	$8 \cdot 10^8$–$6 \cdot 10^{10}$	UV the boundary between ionizing and nonionizing radiation
10^{-2}–10	$3 \cdot 10^5$–$8 \cdot 10^8$	Nonionizing radiation, visible light, and IR
		Microwaves:
10^{-2}–10^{-4}	$3 \cdot 10^4$–$3 \cdot 10^5$	Millimeter waves
10^{-4}–10^{-5}	$3 \cdot 10^4$–$3 \cdot 10^3$	Centimeter waves
10^{-5}–10^{-6}	$3 \cdot 10^3$–$3 \cdot 10^2$	Decimeter waves

misunderstandings concerning the mode and mechanisms of the primary interactions of radiation with living systems.

It was stressed in Chapter 1 that in microwave equipment three basic components may be distinguished: the transmitter (power source), the power transmission and guiding system, and the radiating element (receiver) as shown in Fig. 7. Simplified general remarks concerning the transmitter were also presented. It remains now to consider the second and third elements, also in a very simplified form (a very clear

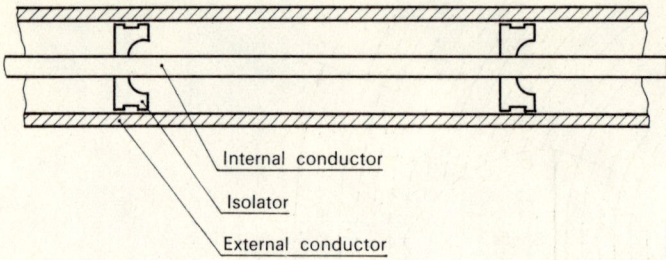

Fig. 14. Coaxial transmission line, schematic cross section. See the text. From [566], by permission.

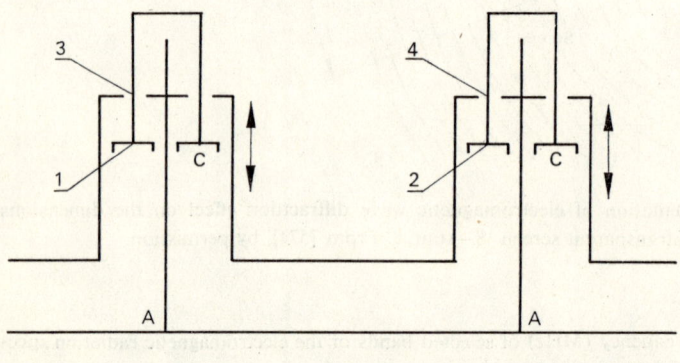

Fig. 15. A double coaxial tuner. To a coaxial line A, two segments of a similar line C are jointed; mobile parts 1 and 2 serve for transmission matching; points 3 and 4 may be the source of nonintended radiation. From [566], by permission.

presentation suited for biologists and physicians may be found in *Lance* [293]). The main hygienic point is of course the problem of intended and nonintended radiation.

Two basic categories of power-distributing systems may be distinguished: transmission lines and waveguides. Transmission lines are used mainly for longer microwaves, i.e., in the decimeter range. The simplest example is the coaxial transmission line (Fig. 14). This is a rigid metal structure consisting of two coaxial conductors separated by insulating material. The external conductor is referred to as the screen. The internal conductor is held in place by dielectric beads or metal supports. In the latter case the distances between the support elements correspond to one fourth of the wavelength. Particular segments of the line may be connected by various coupling elements. The outer conductor of the line the screen eliminates nonintended radiation. Any coupling element may be the point of origin of nonintended radiation, i.e., of hazards from the hygienic point of view or of power losses from the technical point of view. Similarly, all matching elements and attenuators, which serve to optimize power transmission between particular elements (segments), may be the

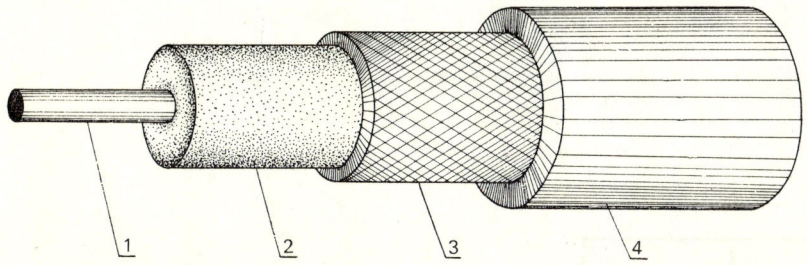

Fig. 16. Coaxial cable: 1, internal conductor; 2, dielectric; 3, external conductor; 4, screen; being of pleated wire strands, the external conductor may be the source of nonintended radiation. From [566], by permisssion.

source of health risks. Other elements that should be considered as potential sources of nonintended radiation are phase shifters, directors, and slotted lines used for measurements. All points of leakage on transmission lines may emit nonintended radiation, thus acting as slot antennas; in most cases, however, irregular distributions of the field arise, with diffraction, reflection, and interference playing an important part in the spatial distribution of the nonintended hazardous field.

In higher-frequency ranges (shorter waves), coaxial transmission lines may be elastic (coaxial cables). The external conductor (screen) is usually made from pleated wires. In this instance slots are always present, constituting a potential source of nonintended radiation (leakage). A coaxial cable is shown in Fig. 16.

Another type of power transmitting element, universally used in microwave technology is referred to as a wave-guide. It is a rectangular or circular hollow metal structure. The propagation of waves occurs on the principle of reflections from the wave-guide walls. This is illustrated in Fig. 17, where the distribution of the E and H vectors is also shown. Note that the wave is retarded in propagation along the wave-guide axis. To pass between points O and O_2, the wave must travel along path $OO_1 + O_1O_2$. If the angle of incidence (equal to the angle of reflection) is designated by α, the path of the wave is

[*11*] $$OO_2 = 2b \operatorname{tg} \alpha$$

where b is the dimension of the wave-guide side in Fig. 17. If this path is divided by time, the velocity of wave propagation within the wave-guide is obtained. The wave travels along path OO_1O_2 with the velocity of light (c), so the velocity of propagation of the wave along the axis of the wave-guide (vg) is obtained from the expression

[*12*] $$vg = c \sin \alpha$$

Thus, as the angle α becomes smaller, the wave travels more slowly along the wave-guide. Further reasoning demonstrates that the possibility of the propagation of a wave within the wave-guide depends on the wave's dimensions. Only a wave that fulfills the condition

[*13*] $$\lambda \leqslant 2b$$

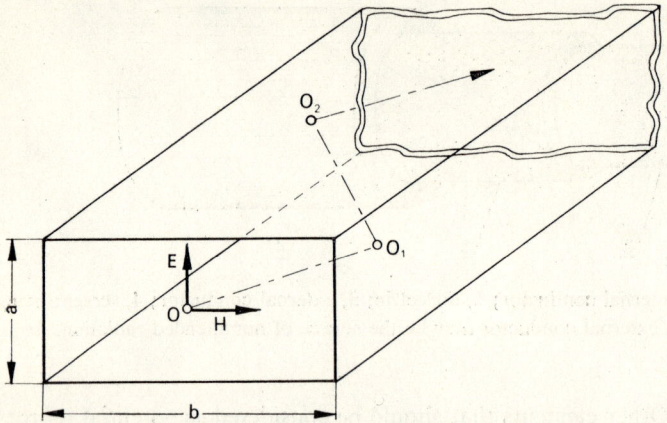

Fig. 17. Wave guide and propagation of an electromagnetic wave along the wave guide, schematic representation. See the text.

where λ is the wavelength and b the distance between the walls of the wave-guide (Fig. 17), may propagate along a given wave-guide. The wavelength equal to 2b is referred to as the critical wavelength. Because the dimensions of the wave-guide and the wavelength must be matched, the use of wave-guides is limited.

Wave-guide segments may be connected to each other by various coupling elements. A wide variety of wave-guide junctions exists. The junction shown in Fig. 18A may be a source of leakage, if the junction along the whole length is not secured. Even if the best leakage-preventing devices are used, the junction must be controlled periodically. Another method is depicted in Fig. 18B, which represents an electric coupling; leakage is prevented by appropriate field distribution. The sector ABC in Fig. 18B may be considered as an open wave-guide at point B and shorted at point C. If the segment ABC is equal to one half-wavelength, the short is shifted to point B. In this manner power may be transmitted from one particular wave-guide segment to another. Displacement of wave-guide segments in point A by more than one tenth of a wavelength may be the source of serious leakage. The same phenomenon may occur if the length of the transmitted waves is changed. Such guiding elements, and various couplings and junctions, directional couplers and phase shifters, attenuators, tuners, and measuring devices may be and often are the source of nonintended radiation. In consequence, unexpected health hazards arise, which are often unaccounted for in hygienic analysis. Directional couplers consist of two wave-guide segments. At the termination of one of these, energy flow takes place. Such an output termination must be linked to a receiver measuring device, load or shorted. If not, the aperture is a source of hazard.

The preceding examples serve only as an illustration of the need for careful analysis of leakage control and prevention in each microwave power guiding and distributing system. One possibility that should be carefully evaluated is the generation

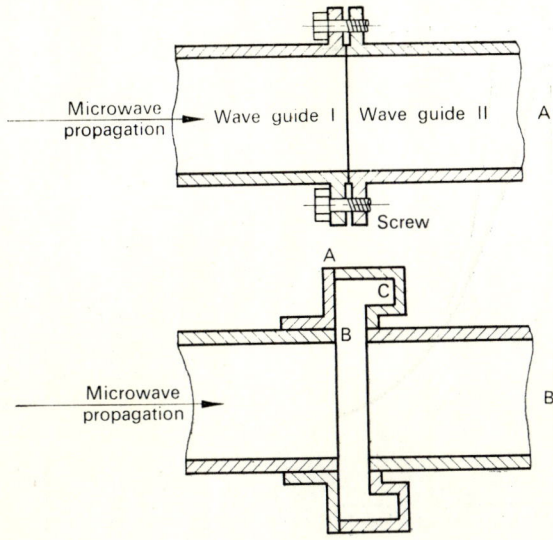

Fig. 18. Wave guide junctions examples. See the text. From [566], by permission.

of additional modes in the transmission system. In the opinion of the present authors, hygienic control must be carried out periodically by a hygienist and an engineer collaborating closely in the prevention of nonintended radiation exposure.

The third element of the schematized microwave equipment is represented by receivers or radiating elements (antennas). Only the latter will be briefly discussed. Receivers, designated for absorption of the total energy transmitted, may be the source of hazards in cases of mismatch. The simplest instance of a radiating element is the dipole antenna, which emits energy in all directions. The amount of energy sent out per unit of time corresponds to the power output. Power per unit of area corresponds to power density (see the concluding remarks to Chapter 1). If the source is isotropic and emits in all directions, the power density P_d at an arbitrary distance R depends on the total power output of the source P_t:

$$[14] \qquad P_d = \frac{P_t}{4\pi R^2}$$

In microwave techniques directional antennas emitting focused beams are used. In this instance, power density depends on the gain of the antenna G, according to the expression

$$[15] \qquad P_d = \frac{P_t \cdot G}{4\pi R^2}$$

This expression is valid only at a certain distance from the source (far-field zone) and along the electric axis of the beam.

38   Physical Characteristics of Microwaves

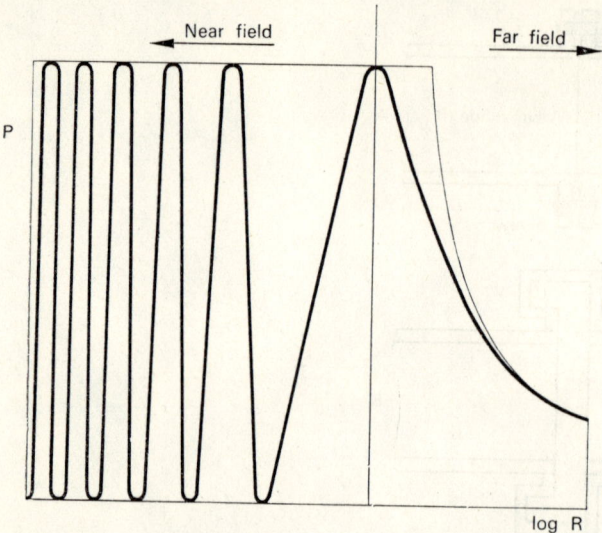

Fig. 19. Power density distribution in the near- and far-field zone, represented schematically as a function of the distance from the antenna. According to [397].

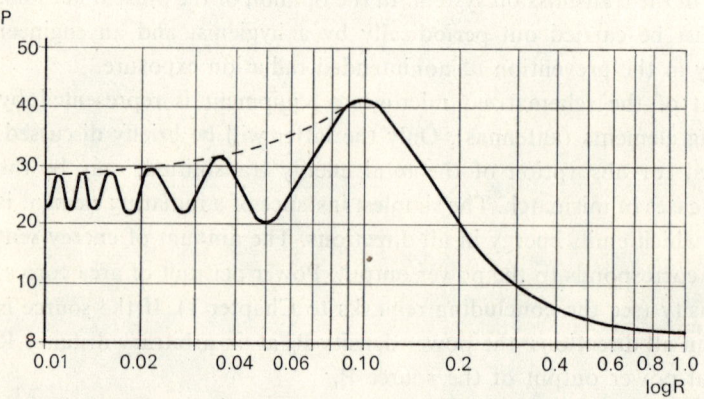

Fig. 20. Power density distribution similar to Fig. 19; oscillations of the power density measured in the near-field zone are presented. From [573], by permission.

In the near-field zone the power density and shape of the beam change according to the dimensions and shape of the aperture of the source (antenna). The value of P (power density) for a circular aperture or a source with determined beam divergence within the near-field zone is shown in Fig. 19. In the far-field zone the power density decreases with the square of distance, according to the preceding expressions. The differences between the power density distributions along the beam axis within the far-field and near-field zones are illustrated in Fig. 20, which is an example of another power density distribution from a similar source as in Fig. 19.

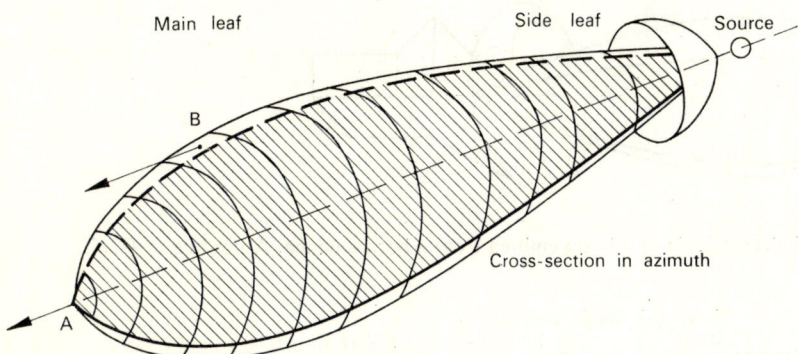

Fig. 21. Power density distribution for a high-gain radar antenna. A and B are reference points to the diagram in Fig. 22. From [574], by permission.

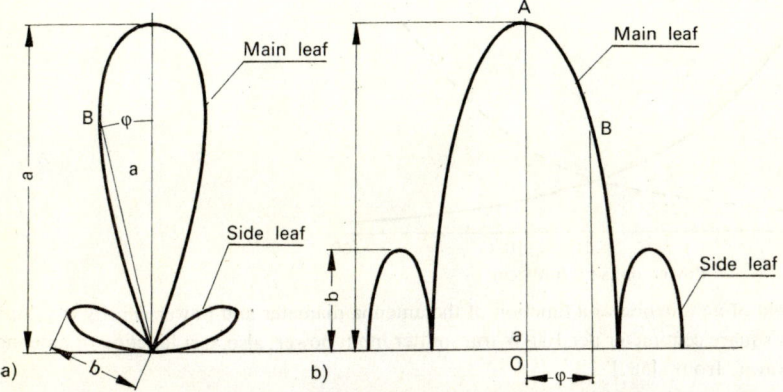

Fig. 22. Power density distribution for the same antenna as in Fig. 21 in polar and rectangular coordinates; compare points A and B. From [574], by permission.

As already mentioned, such representations are simplified and their use for the evaluation of hazards within the near-field zone is questionable. The power density distribution in space may be accepted as a meaningful index for far-field conditions only. At this distance the beam is formed and the directional characteristics of the antenna may be determined. The diagram in Fig. 21 represents characteristic power density distribution for a high-gain antenna, such as is used in radiolocation. In Fig. 22 a similar diagram in polar coordinates and rectangular coordinates (used more frequently) is depicted. The spatial characteristic may be presented also as a horizontal cross section, azimuth characteristic, or in vertical cross section (plane of elevation). From the diagrams of Figs. 21 and 22 it may be deduced that a displacement of the irradiated object (or of the beam) causes a change in exposure; the incident power density varies accordingly. In certain types of antennas not only side

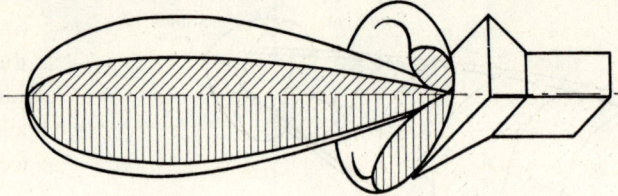

Fig. 23. Spatial characterstic of a beam emitted from a horn antenna. According to [336].

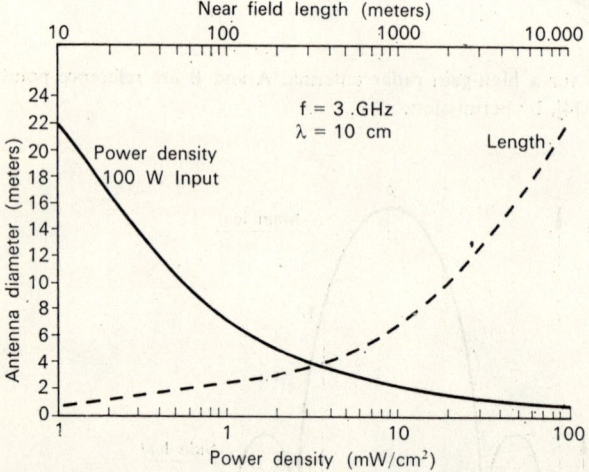

Fig. 24. Near field of an antenna as a function of the antenna diameter and power density envelope in milliwatts per square centimeter per 100-W transmitter input power, also as a function of antenna diameter. Redrawn from [587].

lobes, as depicted here, but also backward lobes may be present. The number of side lobes may also vary. In most instances only the first side lobes are of significance; additional side lobes and backward lobes usually may be neglected in hygienic considerations. To supplement Fig. 22, the spatial characteristic of a beam emitted by a horn antenna is shown in Fig. 23.

Another means of representing graphically the relationship between microwave exposure and antenna size, as well as the antenna characteristics (spatial distribution of power density within the emitted microwave beam), is depicted in Fig. 24, taken from a paper that is very useful for the physician (*Vogelman*) [58].

Various modes for the generation and emission of microwaves exist. The waves may be harmonic; i.e., the electric and magnetic field vectors oscillate according to the law of sines or cosines (sinusoidal waves). The waves may be modulated; i.e., the amplitude, phase, or frequency may be changed in an arbitrary manner. For example in pulse modulation short electromagnetic pulses are sent out at certain time intervals. The duration of the pulse (pulse length, pulse width) is usually of the order of milliseconds and is designated by the Greek letter τ. Its reciprocal or the pulse repetition frequency (pulse repetition rate) is expressed in hertz or cycles per second. The product of pulse length and repetition rate is referred to as the duty cycle. In the case of pulsed-wave generation, the power emitted increases rapidly, reaches a peak pulse power, and rapidly decreases (Fig. 25A). This may be averaged for pulse length (Fig. 25B) or per unit time (Fig. 25C), which introduces the concept of mean (average) power emitted, according to the expression

$$[16] \qquad P_p = \frac{P_a}{f_\Omega \tau} \quad \text{or} \quad P_a = P_p f_\Omega \tau$$

where P_p = peak power,
P_a = average power,
f_Ω = repetition frequency,
τ = pulse length.

In practice, average power is usually measured, and for hygienic purposes mean power density is used. It should be realized that the peak pulse power is many times higher than the average power output. For example, the latter may be on the order of kilowatts, whereas peak power is on the order of megawatts.

In radiolocation, equipment with movable antennas and/or beams (scanning or rotating radars) is used universally, which introduces an additional complication from the medical and hygienic point of view. To the irradiated object microwave power arrives in a series of pulse groups or what could be called pulse packs (Fig.

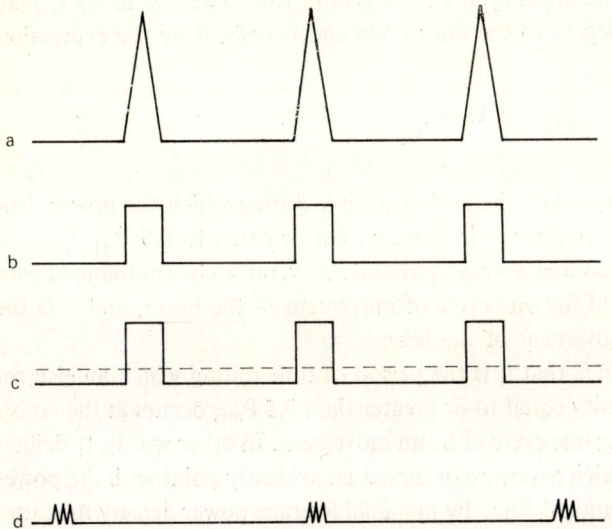

Fig. 25. Schematic representation of pulse modulated wave: a, power peak; b, power averaged for pulse duration; c, power averaged for pulse duration and compared with mean (average) power density average for a series of pulses (dashed line); d, series of pulses emitted at fixed time intervals (illustrates the concept of irradiation cycle rate).

25d). The concept of irradiation cycle rate was introduced by *Deichmann* [116, 120] to designate this mode of irradiation. The concept seems to be very useful in the evaluation of health hazards. Also, in this case the duration of the irradiation by such a pulse series and the irradiation cycle rate should be distinguished. This same aspect of periodical irradiation by a mobile beam can be taken into account in another manner. The concept of effective irradiation time was introduced with this aspect in mind [384]. Effective irradiation time t_{ef} depends on the beam divergency (width) and may be defined by the expression

[17]
$$t_{ef} = \frac{BW}{360°} t_t$$

where BW = beam width,
t_t = total time of power emission.

Unfortunately, this reasoning neglects the irradiation cycle rate, which may be of biological importance. The same objection may be raised to another concept, also introduced in Poland [104]. This concept consists of an arbitrary distinction between stationary and nonstationary fields of irradiation and is used for practical purposes. Stationary fields were defined as fields generated by equipment with a stationary antenna (beam) or a mobile antenna (beam), where the movement is characterized by an index C greater than 0.1. Nonstationary fields are defined as fields generated by equipment with a mobile antenna (beam) where index C is smaller than 0.1 and the irradiation cycle rate is greater than 0.02 Hz. Index C may be compared to a certain degree to the duty cycle and is defined by the expression

[18]
$$C = \frac{t_i}{T}$$

where t_i is the irradiation time, i.e., the period of time during which the power density of the moving or stationary beam is greater than or equal to 0.5 P_{max} (P_{max} is the maximal value of the measured average power density, not to be confounded with peak power). t_i is determined for one cycle of movement of the beam, and T is the duration of one cycle of movement of the beam.

A more literal translation is that t_i is the period of time during which an electromagnetic field of power density equal to or greater than 0.5 P_{max} occurs at the considered irradiated point during one cycle of beam movement. In other words, t_i defines the period of time during which exposure occurs at an arbitrary point with the power density remaining certain limits, defined by maximal average power density measurement.

This concept was adopted as the official interpretation of irradiation (exposure) conditions for hygienic purposes in Poland [597].

To sum up the preceding considerations, it should be stressed that in the case of the irradiation of a biological object the exposure must be analyzed in detail, with the source characteristics and the spatial relationship of the source and target, taken into account. It must be remembered, moreover, that various objects present in the vicinity of the source influence the configuration of the field. Reflections, dispersion and interference may cause standing waves and the presence of nodes (areas of low or high power density). Multipath interference may also lead to a situation in which the power-density concept is insufficient for health-hazard evaluation, see [564]. Target—source coupling should also be considered; the biological target distorts the configuration of the field. The exact quantitation of exposure conditions, i.e., of the source target relationship, is in most real situations extremely difficult. The approximations used for practical purposes are valid for the experimental conditions of exposure in an anechoic chamber, but should be considered as an extreme simplification if applied for analysis of exposure at a work site. Keeping all these restrictions in mind, one can attempt the enumeration of the basic criteria for the hygienic analysis of exposure conditions. Such an analysis should take into account the following:

1. The generation and emission of intended and nonintended radiation.

2. The reciprocal spatial relationship of the source and target under near- and far-field conditions.

3. Irradiation conditions are related to the mode of source generation (continuous- or pulsed-wave generation) as well as exposure to stationary and nonstationary fields.

In the last instance (exposure to nonstationary fields) the irradiation cycle rate may play a role, and it is extremely difficult to decide if it can be neglected, as is usually done. Irradiation in open space or an anechoic chamber may be evaluated for practical hygienic purposes without qualms. But in many instances the industrial hygienist is forced to give his opinion on the admissibility of exposure conditions in rooms with reflecting walls and contents. In these cases the problem is difficult and any standpoint taken is vulnerable.

Microwave sources may be divided into three groups on the basis of their average power output when taking only intended radiation into account [573]:

1. High-power sources over 10 W.
2. Median-power sources, from 0.1 to 10 W.
3. Low-power sources, below 0.1 W.

These three categories were proposed in Poland at the time when conservative safe-exposure limits accepted in 1961; see [460] were binding. If higher levels are accepted the values given could be perhaps one order of magnitude higher.

Radar installations, missile-guiding equipment, certain radiocommunication transmitters, television transmitters (circular emission instead of focused beams), television-microwave links (directional antennas), magnetron measuring stands, klystron pulsed equipment, microwave sources used in nuclear technology, and microwave heating equipment in various industries (the food industry among others) may be cited as examples of high-power sources. Special attention should be drawn to high-power sources in confined space (room with reflecting walls and contents), such as production and assembly lines of magnetron and klystron equipment, transmitters, and radars, as well as repair and technical maintenance shops.

The following types of equipment may be classified as median-power sources: certain types of radar stations, equipment used in radionavigation, microwave television and telephone links, certain kinds of remote guidance and control equipment, many industrial microwave sources, and measuring devices and standard signal generators. Many devices belonging to this last category (measuring devices and signal generators) are low-power sources (low-power klystrons are used).

It should be stressed that similar equipment, such as radars, may have high, median, or even low power output. This is important, because the existence of a radar installation in the neighborhood may cause unwarranted alarm and anxiety. The power density and exposure time at points in the neighborhood may be negligible, and the sole reason for anxiety is that most people associate radar installations with high-power sources. On the other hand, one may observe radars being put in operation on pleasure craft without any reasonable ground for their use at the moment.

The extreme diversity of microwave equipment and its manifold applications do not permit detailed listing. Three particular groups must be mentioned as meriting special attention because of their use:

1. Scientific equipment used for investigations and prototypes; because of atypical and/or new design, unexpected health hazards may arise.

2. Microwave therapeutic (microwave diathermy) or diagnostic equipment used in medicine. The former is used for the intended irradiation of human subjects; the latter belongs almost exclusively to the category of very low power sources.

3. Microwave ovens must be considered a very special category of microwave equipment because of their use in private homes by housewives and children or in restaurants and foodstands, where the technical skill and knowledge of the personnel may be insufficient.

Finally, two principal sources of health hazards should be pointed out: faulty design and equipment in a state of bad repair. Equipment in which a decrease in output power has been noted must be suspected for leakage and, in consequence, considered a nonintended hazardous radiation source. It was stated by *Voss* [590] in his paper on microwave hazard control in design that the cautious Russian standard can be met for industrial microwave systems without any great extra costs. Nevertheless, all technical means of prevention of microwave hazards will be ineffective if technical personnel remove screens and safety devices and short circuit safety locks when putting equipment into operation. As always and everywhere the human factor of light-mindedness and/or the lack of sufficient knowledge are major sources of additional and unnecessary health risks.

It should be stressed that many health-hazard sources can be eliminated by the appropriate design of equipment. This is particularly important if the equipment is intended for the use of persons without technical knowledge e.g., microwave ovens used by housewives. In such cases equipment performance standards should be imposed by law and compliance with such standards strictly controlled. The legislation concerning microwave ovens adopted in the United States may serve as an example of a very satisfactory solution.

CHAPTER 3 | Interaction of Microwaves with Living Systems

The effects of the interaction of microwaves with living systems as observed by the biologist or the physician may be considered for didactic purposes as the end result of three phenomena:

1. The penetration of microwaves into a biological target and their propagation within it.
2. The primary interaction of microwaves with living matter (mechanism of microwaves absorption).
3. The secondary effects induced by the primary interaction.

The term "interaction" is used here on purpose to stress the fact that end results depend not only on the action of microwaves, but are influenced by the reaction of the living system, which is characterized by the property of adaptability, with a large capacity for compensating for the effects induced by external influences. Hence conclusions drawn from model experiments need appropriate corrections. It is the present authors feeling that some physicists and electronic engineers are apt to overlook this fact.

The concepts of physiologic and pathologic compensation may be useful. Physiologic compensation means that the strain imposed by external factors is fully compensated and the organism is able to perform normally. Pathologic compensation means that the imposed strain leads to the appearance of disturbances within the functions of the organism and even structural alterations may result. The borderline between physiologic compensation and the appearance of pathological phenomena is not always easy to determine. This concept is of course a simplification. A much more meaningful concept is *Hans Selye's* theory of stress-adaptation-fatigue-decompensation. The technical reader is referred to *Hans Selye*, The Stress of Life, McGraw-Hill Book Company, New York, 1956.

Penetration and Propagation of Microwaves within a Biological Target

When microwaves arrive at the surface of a biological target, they enter a medium of different dielectric properties, thus affecting the conditions of propagation. Reflection and scattering occur at the surface. Because of the different properties of the biological medium as compared to air, the field intensity, wavelenght, direction of propagation, and plane of polarization may be affected by such phenomena as reflection, difraction, scattering, absorption, interference, and polarization. The biological object consists usually of several layers of complex geometry with various dielectric properties. At the boundary between each consecutive layer the propagation conditions change; reflection, scattering, etc., occur; and the resulting picture is extremely complex. Simplified approaches are a necessity. Certain simplifications in current use may be accepted for the general purpose of evaluating the bioeffects of microwaves. Others are applicable only for the consideration of particular situations and are inadmissible as a basis for generalization. The limits within which a particular concept (approximation) is applicable should be clearly specified; otherwise, serious misunderstandings may arise, particulary in a multidisciplinarian field of investigation such as the bioeffects of microwaves.

The simplest model may be taken as a starting point. A simple harmonic plane wave front enters a single layer of an arbitrary material that has a flat surface. Energy is absorbed as the field is transmitted or coupled into the material, which is assumed to be a lossy dielectric, as all biological media are. Attenuation occurs as energy is absorbed; i.e., the amplitude of the wave decreases. Phase shifts also take place. Both phenomena may be characterized by appropriate constants that depend on the dielectric properties of the medium. Attenuation per unit length is the attenuation constant α defined by the expression

[19] $$\alpha = \omega \left[\frac{\mu \varepsilon'}{2} \left(\sqrt{1 + \left(\frac{\sigma}{\omega \varepsilon'} \right)} - 1 \right) \right]^{1/2}$$

The phase shift per unit length is the phase constant β defined by the expression

[20] $$\beta = \omega \left[\frac{\mu \varepsilon'}{2} \left(\sqrt{1 + \left(\frac{\sigma}{\omega \varepsilon'} \right)} + 1 \right) \right]^{1/2}$$

where in both expressions
- ω = angular frequency,
- ε' = real part of the dielectric constant permittivity of the medium,
- σ = conductivity,
- μ = permeability. As the difference between the permeability of biological media and that of a vacuum μ_0 is less than 0.01 percent, μ may be replaced in these expressions by μ_0 [80]. The properties of the wave and the medium used in these expressions were defined in Chapter 2. The attenuation constatnt is expressed in decibels per unit length and the phase constant in radians per unit length.

The wavelength within the tissue will be considerably reduced (see Chapter 2), which is caused by the high values of the dielectric constants.

The absorbed power density depends on the conductivity and the electric field strength E, according to the expression [253]

[21] $$P = \frac{\sigma}{2} |E|^2$$

Using the attenuation constant, one may define a characteristic depth of penetration d, often called the skin depth of the material as

$$[22] \qquad d = \frac{1}{\alpha}$$

This is the distance at which the power density will decrease by a factor of e^{-2}. At this distance the power density will have decayed to 37 percent of its value at the interior side of the surface. Within this depth 87 percent of the energy will have been dissipated.

Knowing the properties of tissues, neglecting cooling by the blood flow, one can use such an approach for preliminary estimates of energy absorption in biological objects in certain experimental situations [599] and for the selection of the frequency and power levels to be used in the experiment. The applicability of such predictions is very limited. The approach is inadmissible for the interpretation of results in terms of incident versus absorbed energy in a living experimental animal.

A more complex and more realistic model is the plane multilayered object; reflections at layer interfaces are taken into account. The reflection coefficient (e) during transmission of a wave between two media of different complex dielectric constants ε_1 and ε_2 and thickness greater than skin depth is defined by the expression [253]

$$[23] \qquad e = re^{j\varphi} = \frac{\sqrt{\varepsilon_1} - \sqrt{\varepsilon_2}}{\sqrt{\varepsilon_1} + \sqrt{\varepsilon_2}}$$

where j and φ depend on the properties of the interface (see Tables 6 and 7).

A three-layered model is shown in Fig. 26. It consists of skin, fat, and muscles. The layers may have various thicknesses; those of the skin or fat are finite, but the thickness of the muscle layer is assumed to be infinite, because of the reasoning that all microwave energy is absorbed at a certain depth within this layer. This model may be referred to as the semi-infinite-slab model and was investigated in detail by *Schwan et al.* [473, 474, 179, 481, 483, 484] and *Lehmann et al.* [295, 299, 300, 303]. This model may be very useful in solving various problems relative to the medical use of shortwave and microwave diathermy [295, 302, 303]. Its application is restricted, however, by the assumptions concerning exposure conditions and the geometry of the irradiated target. The generalization of conclusions based on this model is inadmissible [253, 493, 564], and in the present authors' opinion its use as a premise in setting up microwave safe-exposure limits is questionable [480, 481] (see Chapter 2).

The applicability of this model depends on the relationship between the wavelength and the radius of curvature of the target surface. The radius must be sufficiently large to allow the assumption that the object is planar. Moreover, the semi-infinite-slab model implies that, besides skin and fat, all remaining tissues have the same electric properties as muscles and form planar layers. The relationship between the wavelength and thickness of consecutive tissue layers as well as their relative geometry (various curvature radii of different tissue layers and body cavities) must be taken into account. Clearly, such a model is inadmissible for the interpretation of results obtained by the irradiation of mice, rats, or rabbits with microwaves a few centimeters or decimeters in wavelength. Conversely, inverse reasoning shows the fallacy of the uncritical transposition of data obtained by small-animal experimentation to larger biological objects, such as men.

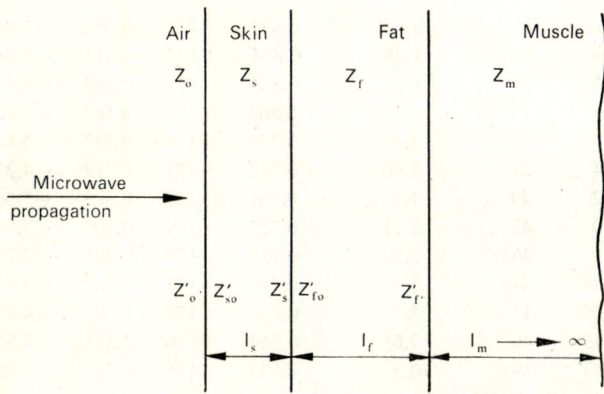

Fig. 26 Three-layered semi-infinite slab model (skin, fat, muscle) of a biological target illuminated by a simple plane wave front. The skin and fat layers have finite thickness I_s and I_f greater than the depth of penetration skin depth; the thickness of the muscle layer is assumed as infinite; I_m, equals infinity, because all the incident energy is assumed as absorbed at this dept. As a whole the model is not transparent for microwaves. Note the interfaces: air-skin, skin-fat, and fat-muscle; because of differences in water content and resultant phase shifts, standing waves may be generated (see the text and Table 6). According to [336].

In other words, this model may be used only if the analysis of the relationship among the wavelength, thickness of consecutive layers, and the geometry of the object as a whole and of particular layers shows the admissibility (sufficient approximation) of this concept. Moreover, it should be kept in mind that the object is illuminated by a simple plane wave; i.e., the reasoning is inapplicable for the near-field zone or complex fields.

If the model is applicable (three-or-more layered), the depth of penetration, reflections, possibility of standing-wave formation, and the distribution of electric field strength in particular layers may be evaluated. Relevant data are presented in Tables 6 and 7.

It should be noted that the dielectric constant and conductivity are frequency dependent both these parameters influence the propagation of electro-magnetic wave through tissues. It should be added that both are also influenced by tem-

Table 6

Characteristics of electromagnetic wave propagation in tissues of high water content represented by muscles and skin at various frequencies; according to [253], slightly modified.

Frequency MHz	Wavelength air/tissue cm	Depth of penetration cm	Dielectric constant	Conductivity mho/cm	Reflection coefficient at interface Air/muscle		Muscle/fat	
100	300/27	6.66	71.7	0.889	0.881	+175	0.650	−7.96
200	150/16.6	4.79	56.5	1.28	0.844	+175	0.612	−8.06
300	100/11.0	3.89	54	1.37	0.825	+175	0.592	−8.14
433	69.3/8.76	3.57	53	1.43	0.803	+175	0.562	−7.06
750	40/5.34	3.18	52	1.54	0.779	+176	0.532	−5.69
915	32.8/4.46	3.04	51	1.60	0.772	+177	0.519	−4.32
1,500	20/2.81	2.42	49	1.77	0.761	+177	0.506	−3.66
2,450	12.2/1.76	1.70	47	2.21	0.754	+177	0.500	−3.88
3,000	10/1.45	1.61	46	2.26	0.751	+178	0.495	−3.20
5,000	6/0.89	0.788	44	3.92	0.749	+177	0.502	−4.95
5,800	5.17/0.775	0.720	43.3	4.73	0.746	+177	0.502	−4.29
8,000	3.75/0.578	0.413	40	7.65	0.744	+176	0.513	−6.65
10,000	3/0.464	0.343	39.9	10.3	0.743	+176	0.518	−5.95

Table 7

Characteristics of electromagnetic wave propagation in tissues of low water content represented by fat and bone at various frequencies; according to [253], slightly modified.

Frequency MHz	Wavelength air/tissue in cm	Depth of penetration cm	Dielectric constant	Conductivity mho/cm	Reflection coefficient at interface Air/fat		Fat/muscle	
100	300/10.6	60.4	7.45	19.1–75.0	0.511	+168	0.650	+172
200	150/59.7	39.2	5.95	25.8–94.2	0.458	+168	0.612	+172
300	100/41	31.1	5.7	31.6–107	0.438	+169	0.592	+172
433	69.3/28.8	26.2	5.6	37.9–118	0.427	+170	0.562	+173
750	40/16.8	23	5.6	49.8–138	0.415	+173	0.532	+174
915	32.8/13.7	17.7	5.6	55.6–147	0.417	+173	0.519	+176
1,500	20/8.41	13.0	5.6	70.8–171	0.412	+174	0.506	+176
2,450	12.2/5.21	11.2	5.5	96.4–213	0.406	+176	0.500	+176
3,000	10/4.25	9.74	5.5	110–234	0.406	+176	0.495	+177
5,000	6/2.63	6.67	5.5	162–309	0.393	+176	0.502	+175
5,900	5.17/2.29	5.24	5.05	186–338	0.388	+176	0.503	+176
8,000	3.75/1.73	4.61	4.7	255–431	0.371	+176	0.513	+173
10,000	3/1.41	3.39	4.5	324–549	0.363	+175	0.518	+174

perature changes [253]; for the microwave region the changes are given by

[24] $$\frac{\Delta\sigma}{\sigma} = 2\% \text{ per } 1°C$$

[25] $$\frac{\Delta\varepsilon}{\varepsilon} = -0.5\% \text{ per } 1°C$$

From Tables 6 and 7 it may be seen that the power absorbed in tissues with high water content may be many times greater than in tissues of low water content (differences in conductivity). If a wave propagating in a tissue of low water content is incident on an interface with a tissue of high water content, the reflected wave is nearly 180° out of phase with the incident wave. A standing wave is produced with the intensity minimum near the interface. In a reversed situation (the incident wave propagates in a tissue of high water content and is reflected at an interface with a tissue of low water content) the amplitude of the reflected wave is in phase with the incident and an intensity maximum is produced near the interface. These phenomena are illustrated in Fig. 27 and may account for subcutaneous burns and thermal necroses observed in experimental animals following irradiation (see Chapter 4).

The standing-wave pattern depends on the thickness of particular layers and wave impedances. The distribution of the electric field intensity and that of the absorbed power density may be calculated; see [253]. These possibilities are best illustrated by the work of *Schwan et al.* (see [487]) and the papers cited previously in the same context. Figure 28, redrawn according to *Schwan* and *Li* [481], is an example of power density absorption patterns as related to the frequency of incident wave

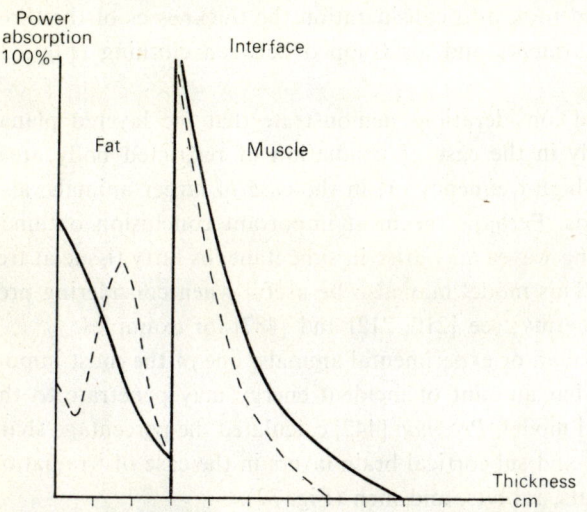

Fig. 27. Relative absorbed power density in two plane layers (fat—muscle) illuminated by a plane wave; note the discontinuity at the interface; continuous line—2450 MHz; dashed line—918 MHz. Redrawn after [253].

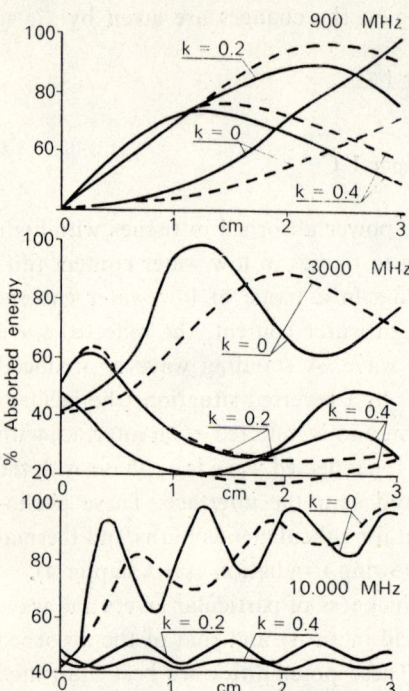

Fig. 28. Relative power density absorption patterns in a plane model at various frequencies for arbitrary skin thickness, k = 0, k = 0.2, and k = 0.4 cm. Continuous line—fat with high water content; dashed line—fat with low water content. According to [481].

and the thickness of subcutaneous fat layer. Figure 29 is another example. For details the reader is referred to the works cited. It may be noted that *Marha et al.* [336] also used this model, and took into consideration the thicknesses of the three layers and the influence of garments and air trapped between clothing (Figs. 30 and 31).

In short, the above-presented considerations demonstrate that the layered planar model may be applied usefully in the case of irradiation of restricted body areas with microwaves in the extra-high-frequency or, in the case of larger animals, also the super-high-frequency bands. Perhaps the most important conclusion obtained using this model is that standing waves may arise in subcutaneous fatty tissue at frequencies higher than 3 GHz. This model may also be useful when considering problems related to medical diathermy; see [210, 212] and [487] for examples.

In the case of irradiation of man or experimental animals, one of the most important points is to determine what amount of incident energy may penetrate to the brain. Using the planar layered model, *Presman* [442] calculated the percentage share of absorbed energy in cortical and subcortical brain layers in the case of irradiation to the back of the head for rats, rabbits, and men (Fig. 32).

From Fig. 32 it could be concluded that only a very small amount of incident energy is absorbed in the brain and only in superficial brain layers. A more realistic model of a lossy sphere used by *Anne* [6] and *Anne et al.* [5, 478] demon-

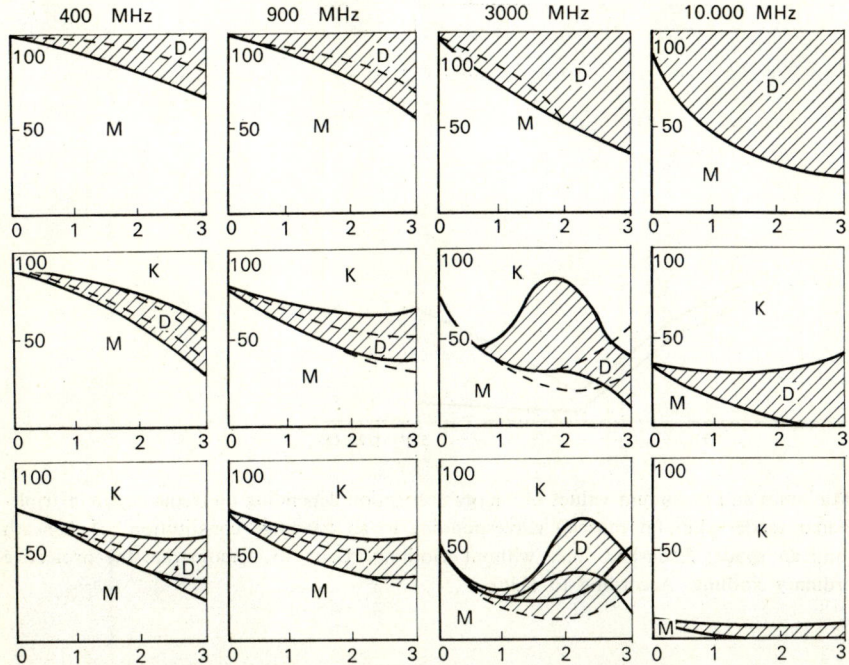

Fig. 29. Relative heating patterns in a planar three-layered model (skin, fat, and muscle) for arbitrary skin thickness 0, top row; 0.2 cm, middle row; 0.4 cm, bottom row as related to the thickness of the fat layer (cm). Continuous line, fat with high water content; dashed line, fat with low water content. K, skin; D, fat, M, muscle tissue. According to [481].

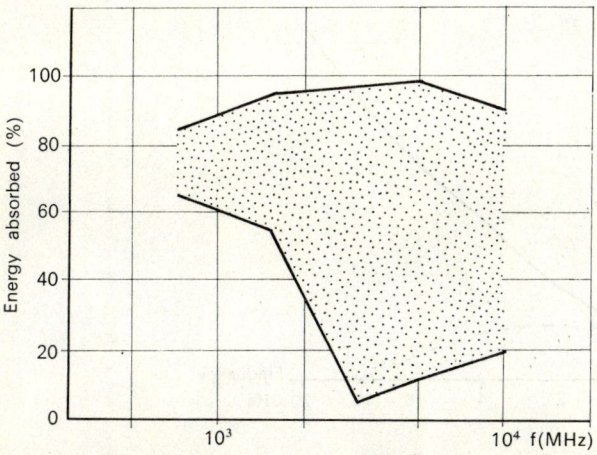

Fig. 30. Maximum and minimum values of energy absorption depending on frequeny in a multi-layered planar model: air, clothing, air, skin, fat, muscle. According to [336].

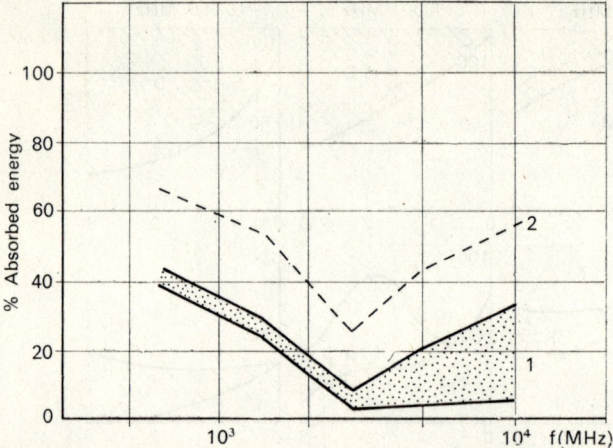

Fig. 31. Minimum and maximum values of energy absorption depending on frequency in a triple-layered planar model (skin, fat, muscle) corresponding to an "average constitution". 1, beneath clothing and air space; 2, dashed line, without clothing. Used to demonstrate the protective role of ordinary clothing. According to [336].

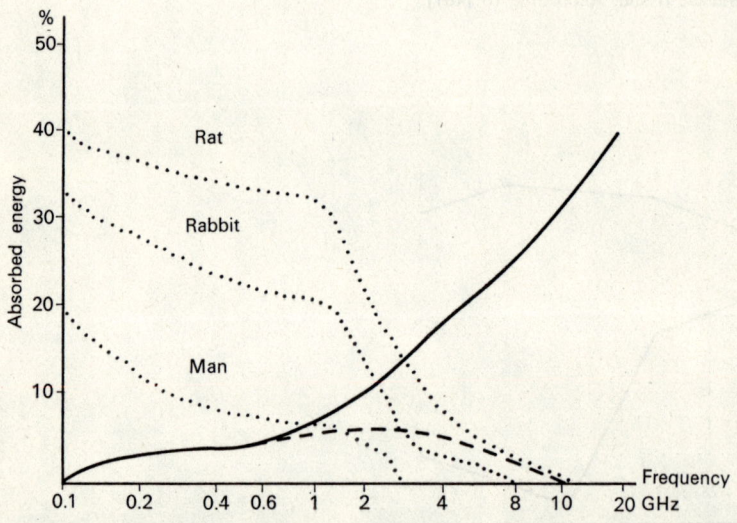

Fig. 32. Percentage share of energy absorbed in the subcortical layers in the rat, rabbit, and man (dotted lines) as compared to absorption in skin (continuous line) and brain cortex (dashed line) depending on frequency. According to [442].

strates that in certain instances the amount of energy absorbed by the head may be far greater than expected on the basis of predictions from a planar model. These authors used the concept of relative absorption cross section i. e., the ratio of the absorbed energy per second to the energy incident on the geometrical cross-sectional area of the sphere, according to the expression

[26]
$$S = \frac{P_a}{P_{di} \pi a^2} = \frac{2 P_a}{E_i^2 \pi a^2 \sqrt{\varepsilon' \mu'}}$$

where S = relative absorption cross section,
P_a = absorbed power,
P_{di} = incident power density (free-space plane wave),
E_i = incident electric field strength,
ε' and μ' = permittivity and permeability of air,
a = radius of the sphere.

It may be seen from this expression that the considered model consists of a homogenous isotropic sphere illuminated by a simple plane wave. From Fig. 33 it may be seen that the amount of absorbed energy varies as a function of the ratio of $2\pi a$ to λ (wavelength), and at certain frequencies may be several times higher than at other frequencies for a sphere of arbitrary size and electrical properties.

According to *Schwan* [474] this sort of macroscopic resonance "is fairly well dampened out by biological fluids and, therefore, not of any great concern." According to this author, the possibility of the existence, at proper excitation frequencies, of hot spots inside the body cannot be excluded. This is not "an urgent task,"

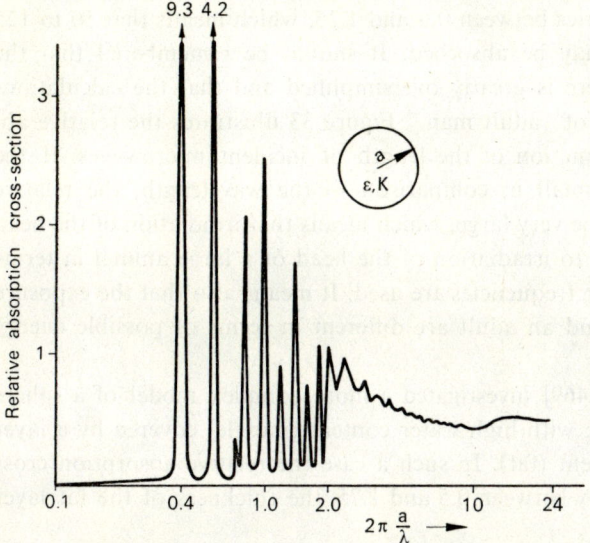

Fig. 33. Relative absorption cross section of a sphere; a, radius; the dielectric constant was assumed as $\varepsilon_1 = 60$; conductivity K = 10 m Mho/cm. Plotted for frequency of 2880 MHz (λ in air). According to [6], redrawn from [474].

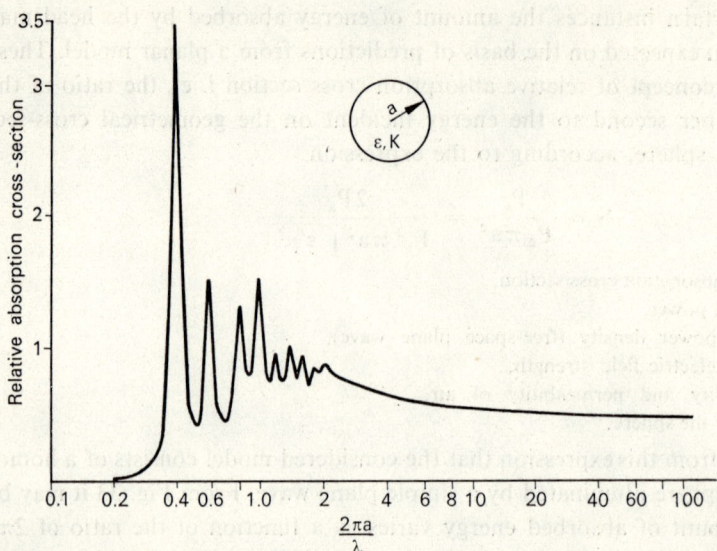

Fig. 34. Relative absorption cross section of a sphere. The dielectric constant was assumed as $\varepsilon_2 = 60$; conductivity $K = 1$ m Mho/cm; a, radius of sphere. Plotted for frequency of 2880 MHz (λ in air). Compare to Fig. 32. According to [6], redrawn from [474].

but more work using similar models with calculations for simple shapes such as spheres and cylinders should be carried out [474].

The general conclusion based on this type of work is that the relative absorption cross section of adult man varies between 0.5 and 1.25, which means that 50 to 125 percent of incident energy may be absorbed. It should be remembered that the model of a homogenous sphere is greatly oversimplified and that the calculations pertain to typical dimensions of "adult man." Figure 33 illustrates the relative absorption cross section as a function of the length of incident microwaves. If the object (sphere) is relatively small in comparison to the wavelength, the relative absorption cross section may be very large, which means that irradiation of the head of a rabbit is not comparable to irradiation of the head of a large animal in terms of absorbed energy, if the same frequencies are used. It means also that the exposure conditions of a small child and an adult are different in terms of possible energy absorption.

Salati, Anne, and *Schwan* [469] investigated a more complex model of a sphere having the properties of tissue with high water content (muscle) covered by a layer of tissue with low water content (fat). In such a case the relative absorption cross section for adult man may vary between 0.5 and 1.75, the thickness of the fat layer playing a very important role.

A much more sophisticated model was used by *Shapiro, Lutomirski,* and *Yura* [493] for evaluation of relative microwave energy absorption (simple plane-wave exposure) in the brain of a monkey *Macaca nemestrina* and in the human brain.

Similarly to *Anne et al.* [5, 6, 469, 478], these authors used the general vector spherical wave solutions of the wave equation, an approach described by *Stratton* [507a]. *Shapiro et al.* [493] presented their solutions in a form amenable to computation by a digital computer. The original papers of *Anne et al.* and *Shapiro et al.* cited here, as well as the review by *Johnson* and *Guy* [253], should be consulted for the equations and the reasoning used. Table 8 presents the values that characterize the model. The mathematics involved are beyond the ability of most physicians so the present authors are satisfied by the fact that the approach of *Shapiro et al.* was endorsed by *Johnson* and *Guy*. From the biomedical point of view, it should be stressed that the model is highly satisfactory from the anatomical point of view and may explain certain empirical observations concerning the effects of microwave irradiation on the animal central nervous system (see Chapter 4).

Table 8
Characteristics of the primate head structure model used by *Shapiro et al.* [493] for investigation of relative microwave energy absorption at 3-GHz frequency.

Region	Tissue modeled	Thickness mm	Radius of surface bounding	Dielectric constant at 37°C	Conductivity at 37°C	Intensity attenuation length mm	Index of refraction
1	Brain	Sphere	26.8±1	42±13	2±0.8	0.08	8.6	6.6
2	Cerebrospinal fluid	2±1	28.8±0.25	77±23	1.9±23	0.02	12.3	8.8
3	Dura	0.5±0.25	29.3±1	45±14	2.5±1	0.11	7.2	6.8
4	Bone	2±1	31.3±0.3	5±1.5	0.2±0.1	0.06	29.8	2.3
5	Fat	0.7±0.3	32±0.5	5±1.5	0.2±0.1	0.06	29.8	2.3
6	Skin	1±0.5	33	45±14	2.5±1	0.11	7.2	6.6

The principal conclusion of this work is that plane models cannot be used for evaluations of the distribution of absorbed power density (called relative heating by *Shapiro et al.*) within the head, and that it may not be safe to assume that radiation of a frequency over 3000 MHz is largely absorbed in the skin, as was postulated by *Schwan* [474, 476]. The results of *Shapiro et al.* indicate that in the animal head peak heating may be expected in the vicinity of the center or around the midbrain region. In the human head model the differences between average heating (power absorption) and the peak value are less pronounced, but nevertheless appreciable. The difference between average and peak absorption is frequency dependent, as is also the exact localization of the peak heating. Figures 35, 36, and 37 illustrate the mean absorbed power density along various-directions, according to *Shapiro et al.* [493]. Some values quoted according to these authors and *Johnson* and *Guy* [253] explain the importance of this approach. An incident field of mean power density of 10 mW/cm² at a frequency of 2450 MHz would produce average heating of about 250 and 600 μcal/g · s average heating as compared to 1000 and 3000 μcal/g · s peak heating in the monkey and human models, respectively. This may be compared to the average resting level of neuronal metabolic heat, i.e., 3000 μcal/g · s [502] as cited in [493]. For the values quoted in Table 8 the maximum relative absorption cross section occurs at 3800 MHz and a neighboring minimum at 1600 MHz. Intense fields occur in the center of the human head and 1.2 cm off center in the animal head at 918 MHz. At the same frequency, absorption at a depth equal to 2.3 times the depth of penetration is two times greater than at the surface. At 2450 MHz, at a depth equal to 4.7 times the depth of penetration, the absorption is 0.43 of that at the surface, which is 5000 times the absorption expected from the plane model. This may serve to illustrate our point about restrictions concerning the use of the model. A discussion of the approach developed by *Shapiro et al.* [493] is given by *Johnson* and *Guy* [253]. The latter authors point out that several hot spots may occur in the brain tissue and are of the opinion that the model for the monkey head may be considered as representative also for cat and rabbit heads. Similar solutions [211, 230] may be applied to models of human limbs approximated as cylinders. In both these papers [211, 230] data may be found on field distribution in tissue layers exposed to the near-zone field from aperture sources. These considerations may be applied to the use of microwaves in medical diathermy, i.e., in controlled situations. It should be noted that the best choice for this use, 750 MHz, is also the most hazardous in an uncontrolled situation [253].

In should be stressed that the spherical model of the head was considered as exposed to a plane wave. It follows that for hygienic evaluation of health risks

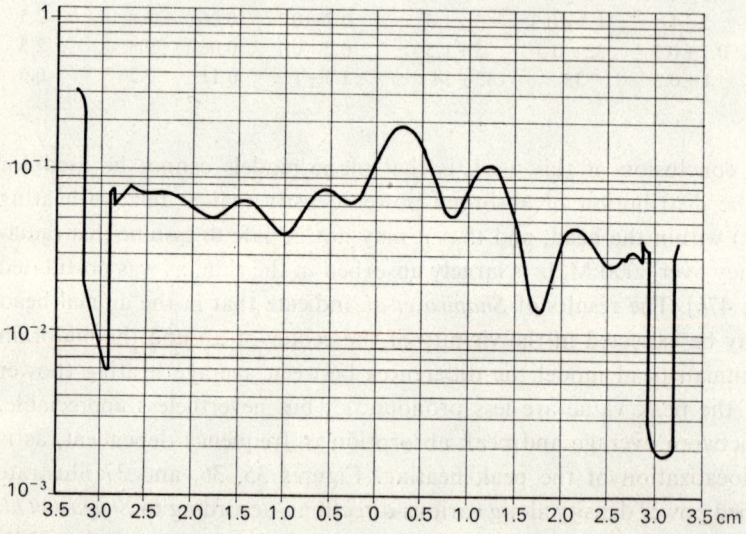

Fig. 35. Mean absorbed power density in a cranial structure along the polar axis. See the text. According to [493].

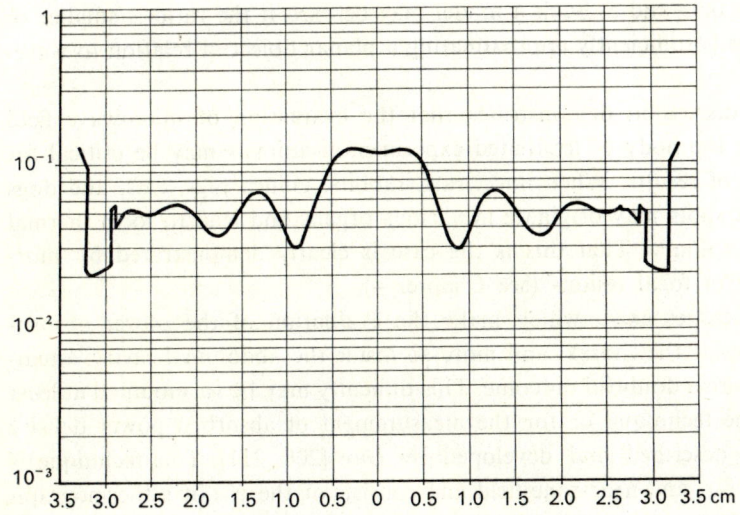

Fig. 36. Mean absorbed power density in a cranial structure in the equatorial plane in the direction of electric polarization. See the text. According to [493].

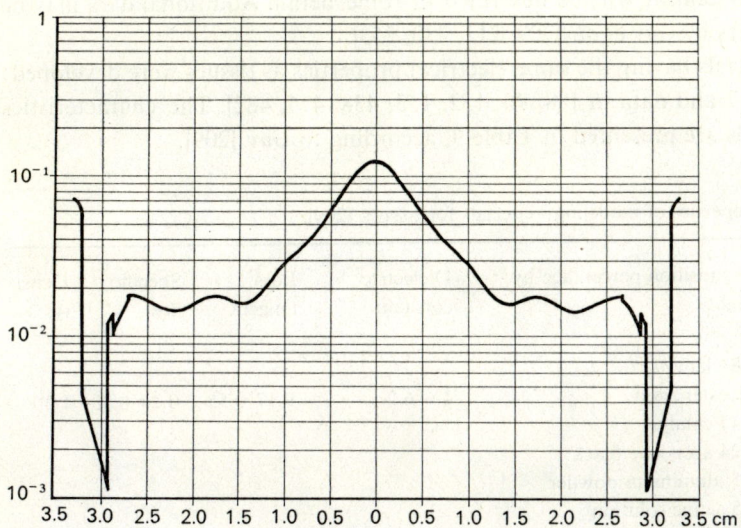

Fig. 37. Mean absorbed power density in a cranial structure in the equatorial plane in the direction of magnetic polarization. See the text. According to [493].

this model is valid for far-field conditions; its value for the evaluation of near-field risks in industry is doubtful. It may serve only as an argument against the admissibility of the semi-infinite-slab model.

As a general conclusion it should be stressed that plane models cannot be used in situations where the ratio between the radii of local curvature and the wavelength lie between 0.05 and 5. Such a model may be used if the ratio is smaller or greater, which means sufficiently approximating a planar object in relation to wavelength.

The foregoing discussion demonstrates that the evaluation of microwave field distribution inside the body of irradiated experimental animals may be critical for the interpretation of results. When mice, rats, rabbits, guinea pigs, cats, and dogs are irradiated, hot spots may originate in various organs and lead to local thermal stimulation and/or injury. That this is the case is clearly demonstrated by morphological finding of focal lesions (see Chapter 4).

Theoretical difficulties exist, which make the evaluation of the power absorption distribution inside the thorax, and more so inside the abdominal cavity, a tedious undertaking with a doubtful outcome. This difficulty may be surmounted at least partly by using the technique of for the measurement of absorbed power density by thermography described and developed by *Guy* [208, 211]. This technique is valid both for the far- and near-zone field and consists of the use of a thermograph camera for recording temperature distribution in an irradiated phantom model or a test animal. Because, at present, this technique is the best simple practical solution of the problem of the determination of the field distribution inside the body irradiated system, it will be described in some detail. Additional data may be found in the papers by *Guy et al.* [209, 211, 230, 253].

Synthetic materials having the same electrical properties as tissues were developed; see Tables 6 and 7 and data in [88, 90, 152, 153, 438, 475, 482]. The characteristics of these materials are presented in Table 9, according to *Guy* [209].

Table 9
Composition and properties of modeling materials for tissues [209].

Modeling material	Composition percentage by weight	Dielectric constant	Loss tangent	Specific heat	Density
Fat and bone, dry	84.81 laminac polyester resin 0.45 catalyst 0.24 acetylene black 14.5 aluminium powder	4.6–6.2	0.17–0.55	0.24–0.30	1.30
Muscle moist jell	76.5 saline solution 12 g salt/liter 15.2 powdered polyethylene 15.2 powdered polyethylene 8.4 "super stuff"*	49–58	0.33–1.7	0.86	1.0
Brain moist jell	63.2 saline solution 9.3 g salt/liter 29.8 powdered polyethylene 7.01 "super stuff"*	33–35	0.27–1.6	0.83	1.0

* Jelling agent available from Whamo Manufacturing Co., 835 East El Monte St., San Gabriel, Calif.

Simulated "body parts" and various tissue structures may be made from these materials and exposed in analogous conditions to the irradiation of test animals. The only difference is that the phantom model should be exposed to a much higher power density to obtain heating in the shortest time possible. In this manner measurable temperature distributions are obtained and may be easily registered by using a thermograph camera. The model may be disassembled along various planes and relative heating patterns determined. A thin 0.0024-mm-thick plastic film should be placed over the precut surfaces to prevent the evaporation of moisture.

This technique may be used to determine the absorbed power density over any internal plane parallel to the electric field in the model. The absorbed power density distribution may be related to the difference in temperature distribution, measured before and after heating with microwaves, by the equation

[27] $$P_a = 4.186 \, pc \, \frac{\Delta T}{t}$$

where P_a = absorbed power density (mW/cm^3),
ΔT = temperature change in degrees centigrade,
t = exposure time in seconds,
c = specific heat of the material,
p = density of the material g/cm^3.

This equation assumes that thermal diffusion may be neglected owing to the short time of exposure. This in turn necessitates compliance with this condition in experimental procedure. Only short exposure times may be used, and the obtained results refer only to the relative spatial distribution of the absorbed energy dose. In other words the technique serves to measure absorbed energy doses and not to determine temperature rises. Absorbed energy doses, their distribution, and the resulting temperature rises may be at least approximately computed for lower exposure levels and longer duration of exposure.

The properties of the "fat tissue" may be varied by adding various amounts of aluminum powder, which permits the simulation of various tissues of low water content. Aluminum powder allows the control of the dielectric constant, and acetylene black, the conductivity. According to *Guy*, the phantom model should be precut along surfaces perpendicular to the tissue interfaces, the surfaces covered by the thin plastic film, and assembled. The time of exposure should be short, 5 to 50 s according to the power density used. Separation of the two halves of the model takes 3 to 5 s and thermographic recording may then be done. The expected theoretical values and experimental data agree well.

Essentially the same technique may be applied to test animals [253]. The animal is sacrificed and frozen in dry ice, or liquid nitrogen in the same position as used for exposure. The body is cast in a polyfoam block and bisected in a plane parallel to the applied source. A thin plastic film is applied to prevent moisture evaporation. The animal is brought to room temperature, irradiated, and thermographic recordings are made, as with phantoms. Irradiation of a living animal of the same species and morphological characteristics in analogous conditions permits the relating of the biomedical observations to the relative power density absorption pattern.

Using this technique on cats, it was demonstrated [253] that the introduction of metal electrodes into the brain (or ECG electrodes for that matter) may change the distribution of the field and cause artifacts induced by local heat stimulation of the region into which the electrode was introduced. Saline-filled glass electrodes may be used. On the basis of these experiments *Johnson* and *Guy* [253] caution:

"No metal electrodes or leads should be used to record physiological data from a region in the animal while it is being exposed to radiation. The metal leads can seriously perturb the field in the tissues and large increases in absorbed power can occur near the termination of the probe... High-resistance wire leads and recording electrodes must be transparent to the fields. Also it is necessary to filter all leads carrying physiological signals and shield the recording instrumentation to prevent the introduction of artifacts into the recording equipment."

Examples are given by *Johnson* and *Guy* [253] from experiments on cats, in which irradiation with 918-MHz microwave produces a similar relative power density absorption pattern as in man.

These remarks are important for the interpretation of the results of animal experimentation reported in the literature. It may be concluded that localized thermal stimulation may be confounded with low-level or "nonthermal" effects. This constitutes one of the most insidious pitfalls in investigations on the biological effects of microwaves.

A very promising approach for the determination of the field distribution inside the body of irradiated animals is offered by the intensive work on the development of implantable probes, which do not cause a significant field distortion. Such attempts are being made in the United States [253]. A small-diameter plastic or glass tube is implanted in the region at which the absorbed power density is to be determined. A small temperature sensor is introduced, the initial temperature recorded, and the temperature rise after a short high-power-density radiation burst determined. Using equation [27], as given for phantom models, the absorbed power density may be determined. In a living animal the whole operation must be performed sufficiently quickly to prevent the cooling effects of the circulating blood to influence the results.

It should be noted, however, that implantable probes, even the best possible from the technical point of view, introduce additional complicating factors from the physiological point of view. In the present authors' opinion, several approaches should be used simultaneously — phantom models, frozen test animals, and implantable probes. It is to be hoped that such probes for the direct measurement of electric or magnetic field components, as well as power density, will be available in the near future.

To sum up may we can make the following statements:

1. Theoretical models of microwave propagation inside biological objects must be very carefully analyzed as to their applicability.

2. Multilayered planar models are admissible only to a very limited extent; their indiscriminate use for predicting biological effects may lead to serious misunderstandings.

3. More sophisticated approaches for approximating the complex geometry of the body shape and of internal organs such as the methods of *Shapiro et al.* [493], are needed to gain insight into the mechanism of the biological effects of microwaves, i.e., multilayered spherical or cylindrical models.

4. Thermography of irradiated phantom models [211] or test animals [253], as proposed by *Guy*, seems to be the best practical solution for wider use at present.

5. Advanced measuring techniques and the development of implantable probes are needed for the better understanding of the biological effects of microwaves; while quantifying microwave fields, many simplifications are made and possible effects often overlooked [564].

Probably much of the reported biomedical findings will need reevaluation in the light of the preceding statements.

Primary Interaction of Microwaves with Living Systems

A living system may be considered at various levels of organization molecular, subcellular, cellular, or in terms of organs systems, or of the whole organism as well as from different points of view in terms of physics (biophysics), biochemistry, physiology, or morphology. In any terms, a living system may be considered as self-regulatory and homeostatic. At our present state of knowledge, it may be assumed that the perturbances caused by microwave irradiation in a living system considered in terms of physics (biophysics) are the underlying factor for the biological effects observed at various levels of organization, when using different methodological approaches. The foregoing sentence defines the meaning of "primary interaction of microwaves with living systems" as used for the purposes of this discussion, and as opposed to the secondary effects discussed in the next section. In view of this definition, the mechanisms of absorption of microwave energy in a living system may be considered as being the main part of the primary interaction.

The results of basic theoretical and experimental investigations on the absorption of microwave and radio-frequency (RF) energy in biological media [88, 91, 474, 479, 481, 484] have been restated and described many times in the literature (for details see the monographs by *Presman* [438] or *Marha et al.* [336] and the reviews by *Cleary* [80] or *Johnson* and *Guy* [253]). Nevertheless, another short, simplified restatement is unavoidable.

Recalling that the permeability of biological media is for practical purposes equal to that of free space, the interaction of microwaves with matter may be considered here as the influence of an external electric field. Energy exchange with free charges or with structures having asymmetrical charge distributions (dipoles) occurs. Charge redistribution under the influence of the field may give rise also to induced dipoles. Dipoles are aligned with the field, and in the case of an alternating fields, changes in orientation lead to rotation with the frequency of the field. The energy required for induction or reorientation of dipoles depends on the dipole moment. The rotation of dipoles is dampened by the frictional resistance of the viscous biological medium; relaxation absorption replaces resonance absorption. These phenomena underlie what is called the thermal effect, i.e., the temperature increase (heating) caused by microwave irradiation.

If the electric field is removed, the aligned dipoles return to their initial state, this phenomenon being characterized as the relaxation time. As *Cleary* [80] expresses it, this is the measure of the time required for the order of the oriented dipoles to be reduced to e^{-1} (37 percent) of its initial value and should be considered a parameter of the dipole structure, depending on the geometry, viscosity, and temperature of the medium. From this it follows that maximum energy absorption will occur if the period of the alternating external field is of the magnitude of the relaxation time (relaxation resonance). Because the relaxation time for water at room temperature is on order of 10 picoseconds, maximum absorption may be expected in the frequency range of 10^3 to 10^5 MHz; i.e., the values of $10 \cdot 10^{-12}$ s are comparable with frequencies of 10^9 to 10^{11} cps (Hz), according to *Moskalenko* as cited by *Marha et al.* [336]. This is in good agreement with experimental data [458].

As mentioned before, the absorption of microwave energy in a medium depends on its dielectric constant and conductivity. These in turn are modified by the applied electromagnetic fields. As it is difficult to improve upon the description of this relationship as given by *Johnson* and *Guy* [253], these authors will be quoted directly:

The action of electromagnetic fields on the tissues produces two types of effects that control the dielectric behaviour. One is the oscillation of free charges or ions and the other the rotation of dipole molecules at the frequency of the applied electromagnetic energy. The first gives rise to conduction currents with associated energy loss due to electrical resistance of the medium, and the other affects the displacement current through the medium with an associated loss, due to viscosity. These effects control the behaviour of the complex relative dielectric constant:

[28]
$$\frac{\varepsilon^*}{\varepsilon_0} = (\varepsilon' - j\varepsilon'')$$

where: ε_0 = permittivity of free space,
ε^* = complex permittivity,
ε' = dielectric constant of the medium, real part,
ε'' = loss factor of the medium or imaginary part.

The effective conductivity σ (due to both conduction currents and dielectric losses) of the medium is related to ε" by:

[29]
$$\varepsilon'' = \frac{\sigma}{\omega \varepsilon_0}$$

and the loss tangent σ is given by

[30]
$$\tan\sigma = \frac{\varepsilon''}{\varepsilon'} = \frac{\sigma}{\omega \varepsilon_0 \varepsilon'}$$

ω = the angular frequency.

The quantity ε* will be dispersive due to various relaxation processes associated with polarization phenomena.

From what was said about the data on the relaxation frequency and dielectric constant of water (high values), it is evident that the dielectric behavior of tissues at microwave frequencies is influenced by the water content. Indeed, the behavior of the dielectric constant and conductivity of tissues of high water content may be, according to *Schwan* [474], expressed by simple equations and predicted for given frequencies, taking into account the weight percentage of macromolecular components and salt content. All these statements are based on the exhaustive researches of this author and his collaborators, cited at the beginning of this section. These papers by *Schwan* [474, 475], *Schwan* and *Piersol* [483, 484], and the monograph by *Presman* (438) should be consulted for further details, quantitative descriptions, and references. Useful data may be found also in the publication by *Tell* [546], with the restriction that only planar layered models are considered by this author.

In connection with the preceding remarks it should be added that *Marha et al.* [336] point out that, because of the shortening of the wavelength in biological media, microwaves may have a wavelength comparable to the dimensions of macromolecules, which may favor direct spatial resonance and result in deformation or damage. At the same time these authors stress the improbability of such an event, because of the high field intensity required for this to occur. *Schwan* [474] emphatically denies the existence of any data indicating "any sort of resonance behaviour" (in the sense of macroscopic resonance). According to *Presman* [440, 443], ionic currents induced by external fields play a role in microwave heating (Joule heat production).

The above-described phenomena explain the mechanism of microwave heating, which aptly is called primary heating by *Schwan* and *Piersol* [483, 484]. The resulting temperature rise and its biological effects will be discussed in the next section. It should be stressed, however, that *Schwan* and *Piersol* distinguish also "specific thermal effects" or structural heating occurring during selective absorption of microwave energy by structures with electrical properties different from those of the surrounding media. According to these authors, such selective heating is possible in biological media only for short periods of time (initial, transient temperature rises during microwave exposure). Heat transfer through blood circulation and conduction does not permit such selective heating, unless special irradiation techniques are used. Moreover, the research conducted by *Schwan et al.* led to the conclusion that the size of a structure to be heated selectively at microwave frequencies must be over 1 mm in diameter.

Before the problem of selective versus volume heating is discussed further, the present authors feel obliged to draw the attention of the reader to an important reservation-according to the subject of this survey, the discussion is limited to phenomena occurring in the microwave range. The phenomena involved in energy absorption and the dispersive behavior of electric properties of tissues at lower frequencies require additional discussion. The reader is encouraged to refer to the monograph by *Presman* [438] and the review by *Cleary* [80], where pertinent references are to be found.

It should be pointed out that selective heating at the level of organs or tissues occurs and was not denied by *Schwan* and *Piersol*. The reservations quoted previously pertain to selective heating at the level of molecular and cellular organization or of isolated groups of cells what could be called "pinpoint heating." All the evidence for the theory of microwave energy absorption and the experiments cited previously seem to speak against such a possibility, at least in the range of biologically significant temperature increases. It should be kept in mind, however, that the methods used may be insufficiently sensitive to discern effects on particular components in a highly complex and highly organized structure, such as the living cell. According to the investigations of *Schwan et al.* proteins demonstrate characteristic frequencies in the range of 40 to 3000 MHz, which they explain by segmental rotation of protein molecules and the consequent effects on total polarization. This may be compared with the characteristic frequency for tissue or cell suspensions of high water content.

As *Schwan* [474] points out, no adequate description for the behavior of electric properties of tissues of low water content, as exemplified by fat, can be offered, owing to a lack of adequate data on the state of water or appropriate mixture formulas. This brings us to the question of bound water in protein molecules or subcellular structures, as raised by *Vogelhut* [588]. The original paper should be consulted for details; the reasoning in a simplified manner may be presented as follows: water bound in macromolecules and subcellular structures (membranes) may be considered similar to doped ice layers, which "melt" owing to microwave energy absorption. Such a change in the bound water layer of macromolecules and subcellular structures must have profound consequences on biological function, affecting biochemical reactions and both passive and active transport phenomena across biological membranes. In other words, the thermal effect of what could be termed a phase change (doped ice structure to free water) was not considered hitherto in evaluating biological effects not attributable to temperature rise. The behavior of the electric properties of proteins and changes in the electrophoretic mobility of irradiated macromolecules fit preceding hypothesis (see original paper). No direct proof for this interesting possibility has as yet been offered.

The hypothesis of *Vogelhut* may be contrasted to the conclusion of *Schwan's* school that the cell interior must be "fluid." This is based on conductivity studies and a consideration of cell-membrane capacity 0.3 to 1 $\mu F/cm^2$ and reactance as compared to the internal impedance of cells at frequencies of 100 MHz and higher. One wonders about the compatibility of such a cellular model with all the data on the highly organized and ordered subcellular structure, as obtained by modern electron microscopy-double-membrane arrangement of smooth and rough cellular reticulum of the Golgi zone and ergastoplasm, nuclear membrane, structure of mitochondria, various fibers, especially muscle fibers, chromosome structure, etc. The excellent reviews by *Cleary* [80] and *Johnson* and *Guy* [253] may serve to stress the point, as demonstrated by a quotation from the latter:

"The tissues are composed of cells encapsulated by thin membranes containing an intracellular fluid composed of various salt ions, polar protein molecules, and polar water molecules. The extracellular fluid has similar concentrations to ions and polar molecules, through some of the elements are different."

To a morphologist, as both present authors are, this cellular model is unacceptable. Being to a certain degree familiar with the precautions needed for the preservation of subcellular structure for electron-microscope study, and the speed of alteration of this structure after removal of tissues from an animal, one cannot but wonder about the applicability to a living, unaltered cell of results obtained during determinations of the electric properties of tissues *in vitro*. Microwave irradiation and all the manipulations during measurements may alter subcellular structure and reflect on the results obtained. This may not be the case; nevertheless, the matter should be investigated, using modern electron-microscopic and cytological methods. No relevant study could be found in the literature surveyed. The present authors consider such carefully conducted studies as absolutely necessary for the acceptance of any explanation of the mechanism of the biological effects of microwaves, and stress the necessity for publication of eventual negative results. It should be stressed that an apparent discrepancy exists between the theoretical explanations available, the experimental results obtained on isolated components (biologically active molecules, membranes), and the effects observed in complex biosystems such as subcellular structures, cells, organs, systems, and whole intact organism. This may be the result of the inadequacy of our knowledge on the biophysics of complex biosystems, or on the primary interaction of microwaves with living matter, or possibly both. This point was raised by *Schwan* during the International Symposium on Biologic Effects and Health Hazards of Microwave Radiation in Warsaw, October 15-18, 1973. Further investigations are needed and more sophisticated approaches must be used.

Another point to be raised concerns the unique structure and metabolism of such a highly specialized cell as the erythrocyte, which was used in many of the studies surveyed. It is difficult to accept generalizations based on such experiments. Membrane structure and the function of such cells as erythrocytes, lymphocytes, and nerve or muscle cells are comparable only to a very restricted degree. In connection with this it should be clearly stated that microwave irradiation may or may not affect directly or indirectly (primary or secondary effects) cellular membrane structure and/or function, and in the opinion of the present authors sufficient data do not exist either as to specific cell types or to an idealized concept of a cellular membrane in general. Without data, such a concept is difficult to admit. It should be clearly stated that the membrane properties of various cells differ; reflecting the differences in membrane structure.

For the record, two views should be presented. *Schwan* and a large group of investigators, who accept his views, are of the opinion that microwave irradiation cannot cause excitation of biological membranes. This standpoint may be presented best by a verbatim quotation from *Schwan* [474], so as to make any distortion of views impossible and to allow comment:

"Membranes are short circuited by currents of high frequency. Straightforward application of Laplace's equation permits the calculation of the potential ΔV evoked across the membrane of a nerve fiber in the presence of the field perpendicular to the direction of its axis. The result is given in Fig. 10 (Fig. 38 in this book). C_M, the capacity of the membrane, has the usual value of about 1 $\mu F/cm^2$ as it is well established in neurophysiological work. Figure 11 (Fig. 39 in this book) shows the frequency dependence of the potential ΔV evoked across the nerve membrane. ΔV is frequency independent at low frequencies and decreases above a certain cut-off frequency. The cut-off frequency f_0 is given in Fig. 11 (Fig. 39 in this book) and is usually smaller than 1 MHz. Introducing typical values of Q_j, Q_a and C one obtains potentials ΔV which are about 10^5 or 10^6 times smaller than the resting potential. According to all modern concepts of neurophysiology about excitation this just cannot stimulate nerves. Let me state it another way. The electrical field strength is something like 500 kV/cm. The field strengths applied by a microwave field to a human body are infinitely smaller and, hence, cannot evoke stimulation."

The following comment should be made. The preceding calculations pertain to normal conditions and no data exist about the influence even to temperature increase on membrane properties. It was shown by *Johnson* and *Guy* [253] that a microwave-induced temperature rise of the thalamus region in the cat below 42°C causes a decrease in latency times of neural responses. *McAffee* [341, 344] described several neurological effects (see Chapter 4) that may be traced back to the influence of temperature increases. Even if the electrical properties of the membrane do not change as a direct effect of microwave irradiation, this does not mean that nerve cell function remains directly or indirectly unaffected. Sufficient data do not exist on the properties of nerve-cell membrane or the nerve cell in various functional states at slightly elevated temperatures. Using the concept of thermal versus non-thermal effects, it cannot be excluded that both may contribute to end effects at cer-

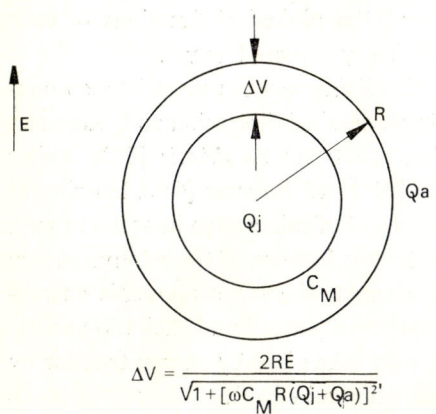

Fig. 38. Alternating potential evoked across a membrane by a field E_0 directed perpendicular to the axis of the membrane of the nerve cell; Q_j and Q_a are specific resistances inside and outside the membrane; C_M, membrane capacitance per square centimeter of surface area; R, cell radius. Redrawn from [474], quoted verbatim.

$$\Delta V = \frac{2RE}{\sqrt{1+[\omega C_M R(Q_j + Q_a)]^2}}$$

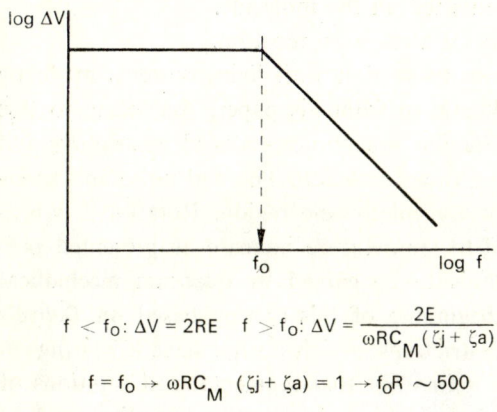

$f < f_0: \Delta V = 2RE \quad f > f_0: \Delta V = \dfrac{2E}{\omega RC_M (\zeta_j + \zeta_a)}$

$f = f_0 \to \omega RC_M (\zeta_j + \zeta_a) = 1 \to f_0 R \sim 500$

Fig. 39. Frequency dependence of ΔV. At low frequencies $f < f_0$; V can be some millivolts for a field $E_0 = 1$ V/cm and, hence, may excite upon appropriate rectification. However, at frequencies $F > f_0$, V is very much smaller and excitation appears impossible. The equation for f_0 indicates a value below 1 MHz, quoted verbatim from [474] and redrawn. For remaining designations, see Fig. 38; ω is angular frequency.

tain energy absorption levels. As *Frey* [165] has pointed out, our knowledge about the mode of transmission, storage, and coding of information in the nervous system is insufficient to take an arbitrary standpoints on possible mechanisms of action of microwaves on the nerve-cell membrane, and the more so on the function of the nerve cell itself.

However, *Lazarev* as cited by *Marha et al.* [336] pointed out that ionic currents induced by electromagnetic fields may change the distribution of ions in the vicinity of cellular membranes, which may affect both their electrical properties and function. Also on this point insufficient evidence exists to admit or to neglect such a possibility.

Another approach to determining possible mechanisms of changes in nerve-cell sensitivity is represented by *Marha et al.* [336]; their monograph should be consulted for details. According to these authors, many parts of the organism may be considered as semiconductors and demonstrate changes in resistance depending on the direction of the current flow and an asymetrically nonlinear volt-ampere characteristic. Rectification of an alternating signal may take place in biological substances, thus affecting both the electrical properties and the biological functions of cells and their membranes. This possibility is insufficiently explored as yet.

The interaction of electromagnetic energy with living systems must be also considered in the light of quantum effects. A pertinent detailed discussion and references may be found in the paper on laser radiation effects by *Vassiliadis* [582]. Other papers to consult are the discussions by *Illinger* [242] and *Presman* [438]. In a simplified, descriptive form the quantum mechanical model of microwave interaction with biological media (living systems) may be presented as follows. If the photon energy of radiation corresponds to quantum differences among the various possible energetic states of a molecule, a quantum of energy may be absorbed and excitation of the molecule occurs. Return to the unexcited state may take place by energy transfer by

1. Increase of kinetic energy of the molecule (thermal effect).
2. Radiation, i.e., emission of photon of lower energy (longer wavelength).
3. Internal motions and rearrangements within the molecule.
4. Providing free energy of activation for a chemical reaction.

The analysis of nonionizing radiation interaction with living systems made by *Illinger* [242] is summarized in Fig. 40, taken from his paper. According to this author, the primary processes in this region are photon-induced absorption and emission and collision-induced absorption and emission. Thermal noise and spontaneous emission may be neglected for the microwave region. Rotational, vibrational, and relaxation phenomena lead to temperature increase as presented previously and no additional information is to be gained by quantum mechanical treatment. The considerations at the beginning of this section based on *Debye's* theory, may be accepted with certain restrictions, which do not have a bearing on biological effects. *Illinger* [242] further points out that segmental coil rotations of biopolymers are probable for radio-frequency and microwave radiations and "...fluctuactions from the equilibrium distribution of tertiary structures are expected..." This of course may influence the metabolic and replicative processes of the cell and cell membrane function; the present authors feel that biophysical and biochemical data are at present insufficient to assess the extent and importance of the biological effects produced.

Perhaps the most important possibilities of primary interaction of microwaves with biopolymers as discussed by *Illinger* [242] are the internal motions and rearrangements within the molecule:

1. Terminal group rotation (bond OH, bond NH_2, and others).
2. Inversion (bond NH_2) and ring deformation.
3. Proton tunneling in H bond bonded systems.

In connection with this it should be recalled that several frequencies in the range from 2000 to 4000 MHz may lead to excitation of ammonia molecules [440]. *Illinger* [242] draws attention also to the quasi rotational or libratory motion of the poly-H_2O sheath as a possible mechanism of microwave absorption, which should be further investigated.

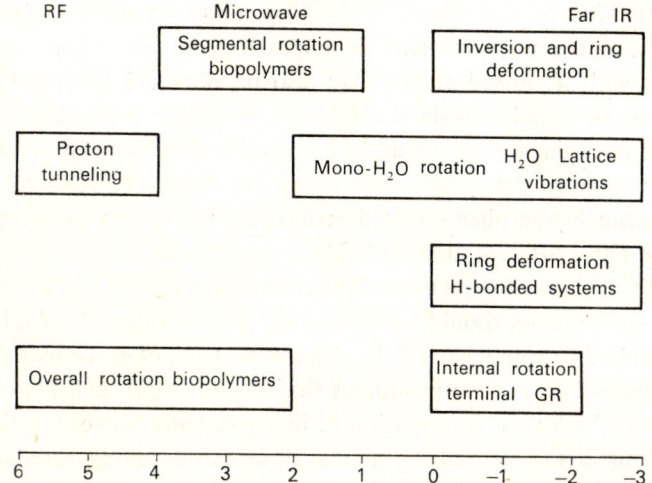

Fig. 40. Expected interactions of nonionizing radiation with biopolymers. Redrawn from [242].

It seems that the quantum-mechanical approach to microwave interaction with living systems is very promising. Resonance absorption phenomena in macromolecules such as proteins [436, 519], methylpalmitate [245], and its related compounds [137], the effects on enzyme activity *in vitro* [436, 438, 440], the effects on the optical activity of glycogen, starch, skin extracts [58, 449, 577] and on electrophoretic motility [13] of proteins or other macromolecules should be investigated further and interpreted using both approaches derived from *Debye's* theory and quantum mechanics. Some effects may be questionable, as for example the electrophoretic changes in human immunoglobulin, for which it was postulated that microwave heating produced the same effects as identical temperature rises obtained by other means [250]. Hemoglobin may be an interesting model because of its properties, well-known structure, and function. Nonlinear effects during transitions of the oxy- and deoxyform *in vitro* evoked by irradiation with 7.1- to 7.6-mm (maximum effect 7.35) microwaves were described recently [290]. At the same time the importance of such research as a tool in the life sciences should not be overlooked — see the interesting discussion by *Jaski* and *Susskind* [252]. It remains only to add that *Tomberg* [558] even envisages ionization when very high power density levels are applied, such as peak power pulses of high-power radars. This is more a theoretical than a practical possibility; nevertheless, it should be kept in mind when the effects of high temperature and high-power microwaves acting simultaneously are encountered.

Another class of phenomena are field-evoked force effects manifested by the alignment of colloidal particles, unicellular organisms, erythrocytes, etc., leading to what is called pearl chain formation. The phenomenon may be explained by the interaction between the external field and dipole (induced dipole) fields of neighboring particles, moving into position, to fit the condition of minimum energy, thus decreasing the initial distortion of the external field. Depending on the electrical properties of the particles and on the external field, pearl chain formation may occur across the field or with it; mixed chains may also be obtained [200, 201]. Pearl chain formation was observed outside the microwave region at frequencies lower than 200 MHz. The phenomenon occurs at field strengths over 100 V/m [474]. In view of this it may be assumed that pearl chain formation does not play a role as a contributory mechanism in the phenomena described in this survey for more details and references, see [200, 201, 224, 336, 438, 474].

As the last of the possible primary interactions of microwaves with living systems, the interesting hypothesis of *Presman* should be mentioned. The monograph of this author [438] must be consulted to appreciate all the arguments. Lack of space makes only a brief description possible. *Presman* postulates that electromagnetic fields of various frequencies transmit biological information in living systems between various component parts. Thus a sort of radiocommunication exists among cellular components within a cell and among various cells within a multicellular organism. Individual organisms are in turn sensitive to magnetostatic, electrostatic, and electromagnetic fields occurring in the natural environment. In view of this, artificial fields may interfere with the synchronization, biological rhythms, etc., of normal functions. The possibility of electromagnetic interference with information transmission by electromagnetic waves within living systems, and between living systems and their environment, awaits experimental verification.

Secondary Effects of Microwave Interaction with Living Systems

The primary interaction evokes a response from the living system, the final result being influenced by many physical and physiological variables. The actual controversy about thermal versus nonthermal biological effects seems to center about the mechanism of the primary interaction of microwaves with living systems; the role of physical and physiological factors is somewhat neglected. In view of this, an attempt will be made to examine the content of the foregoing sections of this chapter using an approach that is customary in radiobiology, which refers to the ionizing radiation part of the electromagnetic radiation spectrum. The biological effects of microwaves may be considered as:

primary interaction: thermal,
                          nonthermal;
secondary effects: late, early (direct, indirect).

The complexity of a living multicellular organism must be once more emphasized. The effects observed at various levels of organization must be analyzed step by step, and consideration given to the limitations and pertinency of the biological methods used for investigations.

The fundamental theoretical and experimental researches of *Schwan* and his collaborators must once more be cited as a starting point. *Schwan* and *Piersol* [483, 484] distinguish thermal, specific thermal, and nonthermal effects. Thermal effects are considered as primary heating and the distribution of thermal energy by condition, radiation, and convection; the blood vessel supply and blood flow play an important role.

When using the approach proposed here, "primary heating" should be considered as primary interaction, and the mechanisms operative in the distribution of the primary heat as factors influencing the secondary early effect. This may be a generalized or local rise in temperature. If the thermoregulatory mechanisms are sufficient to compensate the absorbed energy and dissipate the "primary heat", no temperature changes will be observed. Nevertheless, a thermal effect (secondary) will exist in the form of the activation of thermoregulatory compensatory mechanisms e.g., blood-vessel dilatation and increased blood flow. Thus primary microwave heating may lead to thermal effects:

1. Without a measurable increase in temperature or, more probably, accompanied by transient, slight, local or generalized initial temperature rises.

2. Accompanied by local and/or generalized temperature increases.

The moot point is the possibility of the occurrence of local temperature increases. *Schwan* and *Piersol* define the specific thermal effect as the selective heating of a structure surrounded by media of different dielectric properties. Analysis shows that such a situation does not occur normally in an animal or healthy man. The best example of a specific thermal effect, as defined by *Schwan* and *Piersol*, is the selective heating of a metallic inclusion imbedded in body tissues, which is of course an abnormal situation.

As shown in the first section of this chapter, reflections, scattering, interference, and standing waves may cause local differences in energy absorption, and hence local primary heating. If this attains a sufficiently high level, local thermal injury to blood vessels may occur, with subsequent impairment of heat transfer by blood flow. Foci of localized thermal injury or necrosis are a sufficiently common morphological finding in whole-body, microwave-irradiated animals (see Chapter 4) to postulate that high local differences in temperature are not an exception, but a frequent occurrence. Thermographic findings seem to support this contention. The specific thermal effects seem to be a specific rare instance of a not infrequent phenomenon. From the biological point of view the distinction of this particular instance does not offer any special advantage. The present authors prefer to distinguish between local and generalized thermal effects, accompanied or not by an increase in temperature, and to specify, if needed, the probable underlying mechanism of local effects. The notion of a specific thermal effect will be not used in further discussion.

A thermal effect not involving a rise in temperature seems to be a contradiction. Perhaps it would be better to distinguish between thermal heating and nonheating effects. It is easily imagined that in a given body region energy is absorbed, giving rise to a local temperature increase. If the thermoregulatory mechanisms are sufficiently efficient, the generated heat will be quickly dissipated and no increase in temperature will be observed. Nevertheless, a load is imposed on the thermoregulatory mechanisms, e.g., the blood flow increases. Secondary metabolic effects related to changes in blood flow may occur. If such a situation is repeated, various effects of a secondary nature (late or delayed or repeated chronic exposure effects) are imaginable. A detailed physiological analysis of the subtle effects of such exposures is necessary.

It should be stressed that unequal deep body heating by microwaves is to be expected. The distribution of primary heating depends on wavelength and the external and internal geometry of the multilayered biological object. Hence the distribution of temperatures and thermal injuries, as well as the resulting physiological consequences, are expected to differ in small (mice, rats, medium rabbits, cats), and large (pigs, monkey) animals and in men when irradiated at the same wavelength and power density. It remains to be determined, if corrections for size and wavelength are possible for the formation of more specific general physiological conclusions.

Secondary thermal effects consist of temperature rises. These affect in turn the metabolic rate. Thermal stimulation or the impairment of the function of various organs is to be expected. On general pathophysiological grounds, further consequences affecting the metabolism and/or function of the whole organism are to be expected, if local temperature rises occur in organs exerting controlling functions (nervous system, endocrine, glands, the liver and kidneys). Generalized body temperature increase (hyperpyrexia) will also affect metabolic rates in various organs. If such phenomena occur frequently over a long period of time (chronically repeated single exposures), the physiological consequences may appear as *Selye's* sequence of stress-adaptation-fatigue syndrome. This could be a possible mechanism of the late or cumulative effects of microwave irradiation. The simple heat-balance characteristics of a given animal species exposed to given microwave irradiation (wavelength, field configuration) permit the prediction of the effect of single exposure and the limits of endurance in terms of thermal load. It is to be doubted if the effects of many repeated exposures may be predicted without prior experimental verification.

Local, transient temperature rises may account for the observation of *Subbota* and *Svetlova* that more pronounced effects are obtained by intermittent irradiation than by continuous exposure, when the total energy dose is equal. These authors [517] observed the course of radiation-sickness syndrome (X-rays) in dogs and mice during the regenerative phase in control animals and in animals irradiated with 10-cm pulsed microwaves continuously for 1 h at 153 $\mu W/cm^2$ and intermittently, the total energy dose being the same. The intermittent irradiation was carried out according to the following scheme: 150 $\mu W/cm^2$ for 8 min, interval 10 min, 60 $\mu W/cm^2$ for 8 min, 240 $\mu W/cm^2$ for 6 min, interval 34 min, 320 $\mu W/cm^2$ for 12 min, 60 $\mu W/cm^2$

for 8 min, interval 14 min, 60 μW/cm^2 for 8 min, 150 μW/cm^2 for 8 min. Nervous system and hemodynamic disturbances (EEG and ECG) as well as behavioral effects were observed in the animals irradiated intermittently. The same authors observed that dogs and rabbits adapted to high temperature (hot chamber) or hypoxia collapsed following low-dose microwave irradiation at what is called athermal levels. *Subbota* and *Svetlova* suppose that nervous and endocrine responses to microwave irradiation may influence the resistance of animals to the action of other stressors.

It may be supposed that investigations of endocrine function (hypothalamus-hypophysis-adrenals) and adaptive reaction in animals exposed to repeated low-dose microwave irradiation may contribute to clarification of the "cumulative" or late effects of such exposures.

In view of the preceding considerations, the "thermal-balance approach" to microwave irradiation seems to be insufficient. This approach is based on the following reasoning. Microwave heating is unequal at various depths of the body. Nevertheless, using the semi-infinite, planar, three-layered slab model and assuming temperature equalization by efficient blood-flow cooling, volume heating of the whole organism and its heat-dissipation characteristics are sufficient to predict the reaction and endurance limits. This is a shortcut, which does not take into account the metabolic and functional effects of increased temperature. Moreover, the planar layered model seems to be inadequate (see first section of this chapter). Even if this reasoning is admitted, the strain imposed on the thermoregulatory adaptive mechanisms during repeated exposures remains unaccounted for.

The possibilities of nonthermal effects (rearrangements within macromolecules and subcellular structures) were presented previously, according to *Illinger* [242]. It is obvious that metabolic and functional disturbances at the cellular level are to be expected. Further investigations are necessary to clarify the question. Electron microscopy may prove useful here. Microwave-induced chromosomal aberrations and abnormalities in the structure of cell nuclei and lymphoblastoid transformation of lymphocytes *in vitro* (pp. 133 and 139) are difficult to explain only by primary microwave heating. Starting from considerations on possible nonthermal effects, several suggestions concerning the possible mechanisms for late or cumulative effects could be postulated. Lack of sufficient experimental data does not permit speculation at present, but an open mind should be kept.

The contents of this chapter may be summarized as follows:

1. Improvement of methods for quantitation of complex microwave fields is necessary for the determination of the quantitative relationship between the exposure level and bioeffect.

2. Further advances in investigations of the distribution of microwave fields within the irradiated biological object are necessary for the interpretation of functional disturbances and their relation to absorbed energy.

3. Metabolic effects and effects on subcellular and cellular structure and function should be more closely investigated.

4. Consideration of physiological responses to microwaves in terms of adaptive reactions, stress, and fatigue (desadaptation) may permit more insight into the mechanism of the effects of repeated exposures to microwave irradiation. Table 10, redrawn after *Marha et al.* [336], presents some of the possible mechanisms of microwave interaction with living systems; Table 11 is an attempt to draw attention to the effects at different levels of organization, as investigated by various methods.

Table 10

Probable mechanisms of microwave interaction with living systems; according to [336], slightly modified.

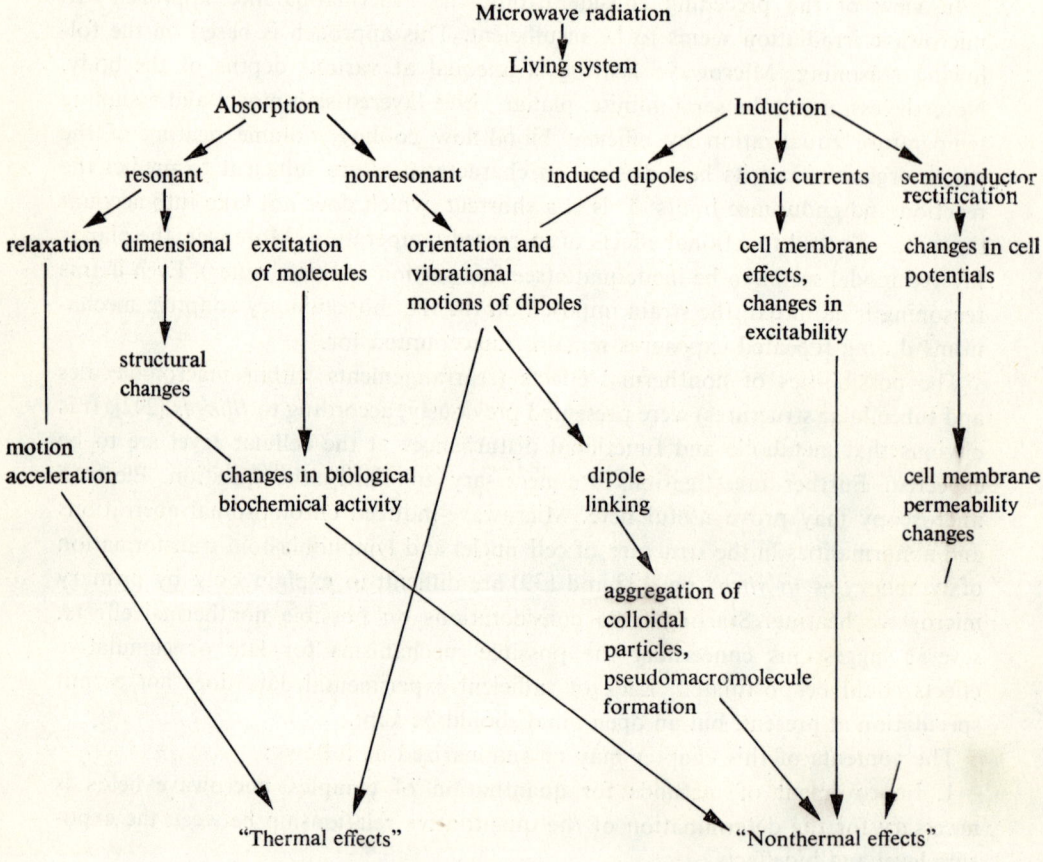

Table 11

Hypothetical schematic representation of microwave interaction with living systems (simplified; see text).

Level of organization	Primary effects	Secondary chain of events	Approaches and/or methods
Molecular	Primary heating, effects on mono- and poly-H_2O, segmental rotation of biopolymers, conformation changes in biopolymers (?), excitation of molecular (?), ionic currents and changes in ion distribution	Temperature rises — metabolic rate effects on biochemical reactions through conformation changes, excitation, changes in ion distribution, etc., structural changes in subcellular elements (chromosomes, mitochondria, membranes)	Approaches developed from Debye's theory and quantum mechanics biochemical and biophysical methods, model experiments
Subcellular and cellular	Interference with biomembranes (or secondary effects through ionic currents and changes in ion distribution), semiconductor effects, changes in bound water	Chromosomal effects, lymphoblastoid transformation, interference with mitosis, genetic effects (?), carcinogenic (?) and/or leukemogenic effects (?)	Classical cytologic methods, cyto- and biochemistry, electronomicroscopy; autoradiography; tissue culture; bacteriologic methods. Largely unexploited
Organs and systems	Focal thermal stimulation and/or lesions	Cardiovascular local and/or generalized effects, nervous system by peripheral receptor stimulation and "disorganization" of function (focal lesions or stimulation), same for endocrine system, effects of changes in metabolic rate and/or abnormal metabolities (?)	Termography [209, 211], morphology, biochemistry, physiology. Determination of energy absorption in the organism
Highly organized living system	Interference with electromagnetic wave transmission and/or reception of biological information (?); see [438]	Stress effects (cumulation of chronic microtrauma and microstress), adaptive responses, desadaptation, interference with biorhythms and their synchronization	As above; electrophysiologic and behavioral methods

CHAPTER 4

Biological Effects of Microwaves. Experimental Data

The aim of this chapter is to present a systematic review of the data found in available literature on experimental facts and observations on the biological effects of microwaves. For clarity's sake the text has been subdivided into sections. The length of each depends to a certain degree on the amount of data available, and does not always reflect the theoretical and/or practical importance of the aspect considered under a particular heading. Therefore, comments expressing personal opinions of the present authors were added. Because of the subjectivity inherent in such comments, the reader is cautioned to evaluate their meaning critically.

Thermal Effects of Microwave Irradiation

The quantitative analysis of all the variables influencing temperature increases in microwave-irradiated animals is far more difficult than in model systems. Because of this, most papers dealing with this subject contain empirical observations; only a few attempts at the quantitative analysis of heat-balance characteristics of microwave-irradiated animals have been made.

The absorption of microwave energy at various wavelengths depends on the spatial, external, and internal geometry of the irradiated object, i.e., the size and structure as well as the position of the animal in relation to the incident planar wavefront [3]. It is difficult to formulate general conclusions, the more so because species specifity of thermoregulatory processes plays an important role. All the data cited on local or generalized temperature increases during exposure at various power density levels are only approximate, and are pertinent only to the determined, actual experimental situation. The data presented in Table 12 illustrate, to a certain degree, the relationship between microwave energy absorption, wavelength, and animal size as evaluated by lethal overheating at various power density levels and various frequencies. At the same time, the table illustrates the restricted value of such comparison. Species specifity of the efficiency of thermoregulatory mechanisms is apparent. The role of this factor is demonstrated even better in Fig. 41. A rabbit weighting 4 kg dies after 30 min irradiation at 2800 MHz and 165 mW/cm^2; a dog of the same body weight and comparable dimensions reacts only with a slight increase

Table 12

Power densities and exposure time until thermal death in various animal species at various frequencies.

Species	Power density mW/cm²	Exposure time min	Frequency MHz	Temperature increase °C	Reference	Remarks
Dog	165	270	2,800	4–6	[149]	
Dog	330	15	200	5	[3]	LD₅₀
Dog	220	21	200	4	[3]	LD₅₀
Rabbit	300	25	2,800	6–7.5	[149]	
Rabbit	165	30	200	6–7	[150]	
Rabbit	100	103	3,000	4–5	[185]	
Rat	300	15	3,000	8–10	[185]	
Rat	100	25	3,000	6–7	[185]	
Rat	40	90	3,000		[185]	
Rat	400	13–14	10,000	7	[394]	
Rat	300	15	24,000	5	[118]	

in rectal temperature. It may be stated in a very general way that exposure of laboratory animals at high power densities, over 100 mW/cm², causes after a relatively short time interval, depending on size and species, a critical rise in rectal temperature and death [3, 23, 25, 122, 149, 150, 195, 294, 318, 356, 357, 364, 394, 553]. At lower power density the time needed to reach critical (lethal) temperature levels is correspondingly longer. In practice, acute lethal effects may be obtained at power densities below 100 mW/cm² only in small laboratory animals; lethal effects below 10 mW/cm² are possible in mice. These animals are notoriously sensitive to ambient temperature increases, which points to a low efficiency of thermoregulatory mecha-

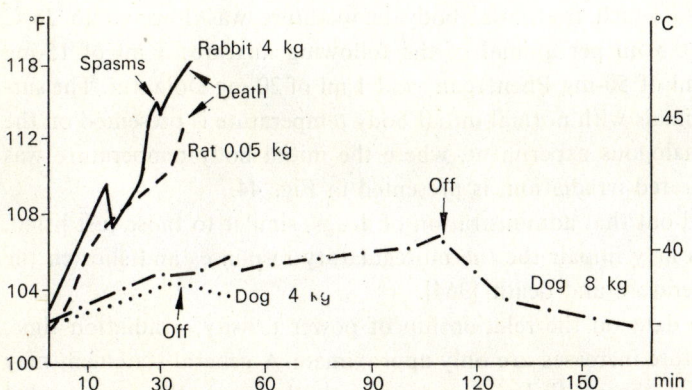

Fig. 41. Thermal responses (rectal temperature increases) in a dog, rabbit, and rat during microwave irradiation (2800 MHz, 165 mW/cm²). Redrawn according to [352].

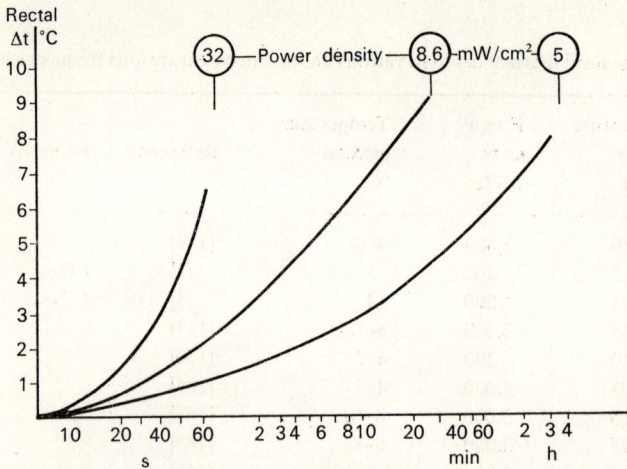

Fig. 42. Rectal temperature increases in mice irradiated with 3-cm microwaves till death. According to [25].

nisms. It should be noted that at low power densities the lethal rectal temperature attained after a prolonged irradiation (Fig. 42) is higher, up to 45°C, than the critical temperature noted during exposures to high power densities (42 to 44°C in mice and rats and 41 to 42°C in dogs, rabbits and guinea pigs). A possible explanation is that the longer time interval before death allows better temperature equilibration among various parts of the body. This observation reflects the fact that microwave irradiation causes inhomogenous body heating.

The preceding data pertain to normal animals irradiated in standard laboratory conditions. Prolongation or shortening of the survival time of irradiated animals may be obtained by altering the initial body temperature [23, 25]. Figure 43 shows the temperature increase and survival time in microwave irradiated mice (3 cm, pulsed, 32 mW/cm²) in which the initial body temperature was lowered to 26°C by administration of 0.5 ml per animal of the following mixture: 3 ml of 15-mg Chloropromazine, 2 ml of 50-mg Phenergan, and 1 ml of 20-mg Dolantin. The survival time curve of animals with normal initial body temperature is presented on the left in Fig. 43. An analogous experiment, where the initial body temperature was increased by prior infrared irradiation, is presented in Fig. 44.

It should be pointed out that administration of drugs, similar to those just listed, or anesthetics to dogs may impair the thermoregulatory responses and shorten the time till critical temperature and death [364].

As said before, the data on the relationship of power density, irradiation time, and resulting temperature increases are only approximate. A general statement may be made that the lower the applied power density, the longer is the time needed for attaining critical temperature i.e., survival time. This is illustrated in Table 13, which shows data obtained from irradiation with 24,000-MHz [122] and 10,000-MHz

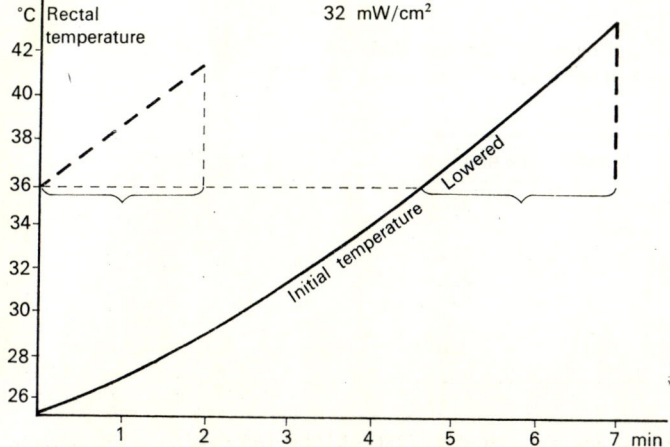

Fig. 43. Survival time and rectal temperature increase in white mice irradiated with 3-cm microwaves after lowering initial body temperature to 26°C, compared with controls top left (dashed line). According to [23].

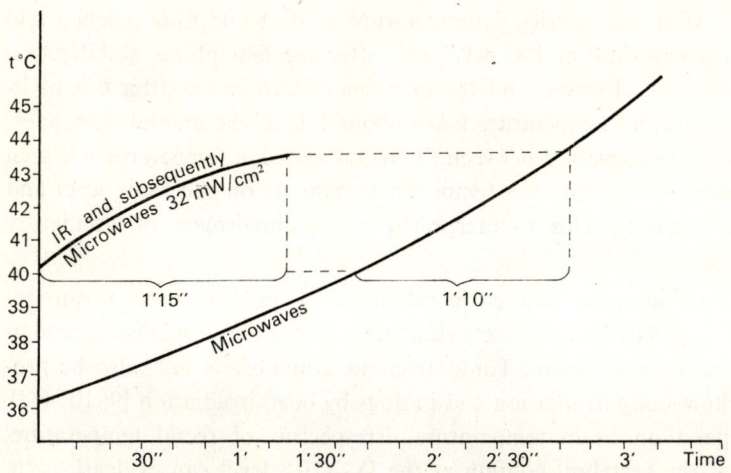

Fig. 44. Survival time and rectal body temperature increase in mice heated with infrared and subsequently irradiated with 3-cm microwaves top left, compared with controls. According to [23].

[115] microwaves. These data and those in Table 12 refer to standard laboratory conditions. Changes in ambient temperature, air humidity, and movement influence heat dissipation and, in consequence, the survival time [112-114, 118, 120, 123, 353, 356, 364]. Although environmental conditions influence the time course of the reaction to microwave irradiation, the basic phenomena are similar. These were investigated in detail in dogs and rabbits irradiated with 200, 1250, and 2800 MHz and at 100 and 165 mW/cm^2 [2, 237, 239, 356, 358, 359, 364, 490]. Three distinct

Table 13

Power density and survival time of mice and rats irradiated with 24,000-MHz (1.25-cm) and 10,000-MHz (3-cm) microwaves.

Power density mW/cm^2	Survival time min	
	Rats	Mice
180	33	5
80	56	13
50	80	35
30	135	140

phases of reaction to thermal acute irradiation (2800 MHz, 165 mW/cm^2, 30 percent air humidity) may be discerned in dogs. The first phase lasts for about 30 min and is characterized by a rectal temperature increase of 1 to 1.5°C. The second phase of relative thermal equilibrium lasts for about 1 h. The third phase is characterized by a rapid temperature increase because of the thermoregulatory insufficiency ensuing from overload. After this a critical temperature of 42 to 43°C is reached and death occurs. During exposure at 100 mW/cm^2, after the first phase, stabilization of rectal temperature was observed and the experiment terminated after 6 h of irradiation. Return to normal temperature takes about 1 h. If the animal is in anesthesia, equilibration of temperature between the rectal and skin temperatures is seen during the first phase. Afterward skin temperature remains on the same level and rectal temperature increases. This is interpreted as an impairment of circulatory responses by anesthesia.

A rapid temperature increase is noted in rabbits at 165 mW/cm^2, and a transient drop may be noted (Fig. 41). In a relatively short time convulsions and death, accompanied by a temperature rise, ensue. Tonic or clonic convulsions may also be provoked in rats by whole-body irradiation and in dogs by head irradiation [9, 10, 485]. This seems to depend on brain temperature, irrespective of rectal temperature. Local irradiation of the vertebral column at the IV-XIIth level causes death without prior convulsions. Death with convulsions may be obtained by head irradiation in monkeys [12] and rabbits [409].

The importance of the relationship between the size of the irradiated object and the wavelength was demonstrated by investigations of the survival time of rats exposed to power densities of 100, 40, and 10 mW/cm^2 at various microwave frequencies, millimeter, 3-cm, 10-cm, and decimeter waves [185, 318]. All animals irradiated with 10-cm waves at 100 mW/cm^2 succumbed within 1 h; at other frequencies about 50 percent succumbed within 1 h. Five phases could be distinguished in the behavior of rats irradiated at 100 mW/cm^2:

1. The animals become uneasy, and cleanse the snout and fur (6 to 10 min).

2. The uneasiness becomes more pronounced; feet, tail, ears, and snout redden (10 to 17 min).

3. Hyperemia of the skin increases; the animals fling about in the cage and make attempts at escape; animals in anesthesia waken (17 to 27 min).

4. Marked skin hyperemia and prostration are observed; sometimes clonic convulsions and edema of the head and genitals appear the twenty-seventh and following minutes; convulsions about the fiftieth minute.

5. Final phase: the animals lie on side, serous exudate appears at the snout mouth, nose, death occurs about the sixtieth minute; no burns are seen.

In connection with the observation that anesthetized sleeping animals waken, it should be pointed out that analeptic effects of microwaves during irradiation of the head or extremities were observed in anesthetized dogs, rabbits, cats, and rats. This effect is discussed in detail by *McAfee* [340, 342, 345, 347], who explains it by microwave heating and the resultant stimulation of peripheral nerves.

These and other investigations on temperature increases induced by microwave irradiation may be summarized by the following general statements [36, 52, 151, 187, 190, 480, 515, 553]:

1. About 40 percent of microwave incident energy is absorbed and transformed into heat at wavelengths below 10 and above 30 cm; absorption within the 10- to 30-cm wavelength range varies greatly; even 100 percent absorption may occur [151, 394, 480].

2. Quantitation of the heat-balance characteristic is difficult in microwave-irradiated animals, although certain formulas have been proposed and suggestions advanced to use such a model for investigations on thermoregulation in animals, for details see [246]; inhomogeneity of the heating seems, however, to offer serious difficulties in the correct estimation of heat-balance characteristics of irradiated animals.

3. Power densities over 100 mW/cm^2 should be considered as a high level of exposure, causing death from overheating in the majority of laboratory animals in a relatively short time; for dogs this value is higher and amounts to about 165 mW/cm^2.

4. Power densities of 10 to 100 mW/cm^2 caused varied body rectal temperature increases in various species, such as rabbits, guinea pigs, rats, and mice; in small laboratory animals thermal death may be obtained within this power density range; no lethal effects are obtained, mice excepting, at 10 mW/cm^2 [23, 25]; effects within the discussed power density range are illustrated in Table 14 which contains the survival times of irradiated rats based on the monograph by *Gordon* [185].

5. In principle, exposures at power density levels below 10 mW/cm^2 do not lead to body rectal temperature increases exceeding the compensatory possibilities of thermoregulatory mechanisms, even in small laboratory animals; within the range of 5 to 10 mW/cm^2 insignificant temperature increases are observed. No measurable temperature increases occur in normal conditions during exposures at power densi-

Table 14

Survival time of rats at various wavelengths and power densities.

Microwave band	Survival time at power density		
	100 mW/cm^2	40 mW/cm^2	10 mW/cm^2
Decimeter	30 min	120 min	More than 5 h
10 cm	5 min	30 min	More than 5 h
3 cm	80	180 min	More than 5 h
Millimeter	120	180 min	More than 5 h

ties below 5 mW/cm^2; 1 mW/cm^2 may be considered the "athermal" exposure level, even at higher ambient temperature and air humidity [398].

The preceding statements refer to single exposure of animals to stationary (see Chapter 2) microwave fields; the thermal aspects of human exposure are presented in Chapter 6. A few more restrictions must be added. Certain authors [1, 21, 101, 185] draw attention to different results obtained by continuous or pulsed wave irradiation. *Gordon* [185] observed shorter survival times in rats irradiated with 10-cm pulsed waves than by CW at identical mean power densities. *Abrikosov* [1] obtained contrary results. If one assumes that the survival time of irradiated animals i.e., till the moment at which critical temperature is reached is influenced by various physiological factors, e.g., peripheral nerve stimulation as manifested by the analeptic effect (see [342]), differences in the action of pulsed-wave (PW) or CW microwaves may be imagined. The amount of heat generated at the same power density level by PW or CW microwave absorption is the same. The time course of heat generation and dissipation may be sufficiently different to influence physiological responses, and the end result e.g., peripheral nerve stimulation may be different. If this were the case, pulse width and repetition rate may play a role irrespective of mean power density and wavelength. It is to be regretted that no systematic investigations of such a possibility have been carried out. Differences in the effects of low-dose chronic PW and CW microwave irradiation at identical mean power densities on the nervous and hemopoietic systems exist, see pp. 104-105 and 144-146.

The role of physiologic factors in the thermal reaction of the organism temperature increase is manifested by adaptation to successive exposure. A decrease in observed body rectal temperatures on successive exposure to the same power density was observed by *Deichmann* [114, 116, 121, 122]. From these observations this author introduced the concept of irradiation cycle rate [120]. Irradiation of rats at 300 mW/cm^2 with 24,000-MHz pulsed microwaves (pulse width 0.6 μs) without interruption (stationary field) leads to death within 15 min [120]. Intermittent irradiation, i.e., corresponding to irradiation with nonstationary fields with a moving antenna or beam, prolongs the survival time of irradiated animals, defined here as the actual time of exposure to microwaves (effective irradiation time). This obser-

Table 15

Survival time of rats during intermittent exposure to 24,000-MHz microwaves at 300 mW/cm² depending on the relationship between duration of exposure period and duration of intervals between exposures (on-off periods) according to [118], selected values.

Operation period of the transmitter (s)		Relationship between duration of exposure and duration of periods between exposures	Survival time equal to effective irradiation time (min)
On	Off		
60	60	1:1	16.5
15	15		28
3	3		40
30	60	1:2	39
10	20		65
3	6		95
60	180	1:3	28
10	30		76
3	9		110 to 120
30	120	1:4	70 to 75
15	60		Over 100

vation is illustrated in Table 15, based on *Deichmann et al.* [120]. This observation was confirmed by *Kowatch et al.* [284]. In these experiments, the duration of the interval between succesive irradiation periods is important. During this interval, no additional energy is absorbed and the animal may dissipate the heat into the surrounding environment. Nevertheless, these observations should be considered together with experiments of *Michaelson et al.* [359], who observed decreasing rectal temperatures in dogs exposed to identical conditions 5 days a week during 4 weeks (1285 MHz, 100 mW/cm²). Each week during successive days the rectal temperature was lower on termination of the irradiation than on the preceding day. After irradiation on the first day of successive weeks, the temperature was higher than on the last day of the preceding week, i.e., after an interval of 2 days without irradiation. The thermal reaction during the fourth week is insignificant when compared with the reaction during the first week. This is illustrated in Fig. 45.

The preceding observations indicate that thermoregulatory mechanisms may show adaptation to repeated thermal loads caused by microwave irradiation. These data were confirmed by Soviet authors [514, 515, 552] on other animal species.

In spite of the objections that may be raised, attempts at quantitation of thermal-balance characteristics of microwave-irradiated animals may be cited, with species differences being taken into account [232]. This is of course only a first appro-

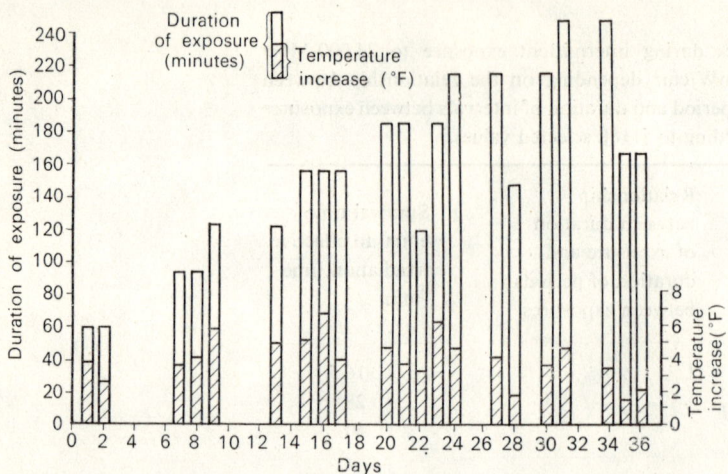

Fig. 45. Rectal tempertaure increases in dogs on successive exposure to microwave irradiation 2800 MHz, 165 mW/cm². Redrawn from [353].

ximation. A theoretical model is considered; a determined body mass G, a surface S and specific heat C_B are used. Metabolic heat M and the heat generated as the result of microwave energy absorption E are the sources of heat in this system. Heat dissipation is calculated from the difference between the temperatures of body surface and air. According to the author cited, this expresses in a relatively accurate manner the heat dissipated by conduction, but not the radiated heat (see also [241]) Specific thermoregulatory mechanisms, such as evaporation from the tongue in a dog and respiratory rate acceleration, are not taken into account. A correction for air humidity and movement should also be introduced (see Chapter 6); in man the role of clothing is an important factor. By admitting these simplified assumptions, the following expression may be used:

[31]
$$t = \frac{GC_B \Delta T}{E + M - S_\alpha(\varphi + \Delta T)}$$

where t = time necessary to obtain body temperature increase by T,
 α = index of heat dissipation by exchange between body surface and surrounding air,
 φ = initial difference in temperature between body surface and air, for the remaining symbols, see the text.

The energy of the absorbed microwaves may be approximated as:

[32]
$$E = \frac{G^I I}{J}$$

where G^I = effective cross section,
 I = radiation intensity,
 J = Joule's index.

The expression is based on the assumption that nearly all incident energy is absorbed. Standard body weights and surfaces may be found in various biological data handbooks. The cited author used data from the handbook edited by *Spector* [505] and presented in Table 16. Certain values may be questioned. The metabolic heat was calculated from the basal metabolic rate, doubling its value. One third of the body surface was used as the effective cross section, which also may be questioned. The specific heat was assumed to be equal to 0.83 kcal/kg, according to *Brody* [54], and the heat dissipation index as 10 kcal /m²/°C/h, according to *Herrington* [227]. The latter paper contains many useful data for similar considerations.

Table 16
Mass, body surface, and metabolic heat in various animal species (data from [505], cited according to [232]).

Species	Man	Cat	Dog	Guinea pig	Monkey	Mouse	Rabbit	Rat
Mass kg	65	3.0	15.0	0.8	3.2	0.02	3.5	0.2
Body surface m²	1.83	0.2	0.65	0.071	0.26	0.005	0.2	0.031
Basal metabolic rate kcal/m²/day	910	750	800	675	610	525	810	905
Metabolic heat kcal/s	3.84×10^{-2}	3.46×10^{-3}	1.2×10^{-2}	1.01×10^{-3}	3.56×10^{-3}	7.86×10^{-3}	3.82×10^{-2}	6.48×10^{-4}

Starting from the preceding assumptions, curves representing the relationship between the power density and exposure time necessary for obtaining a rise in body temperature by 5°C may be drawn, with the additional assumption that the initial difference between the temperatures of body surface and air is 10°C (Fig 46).

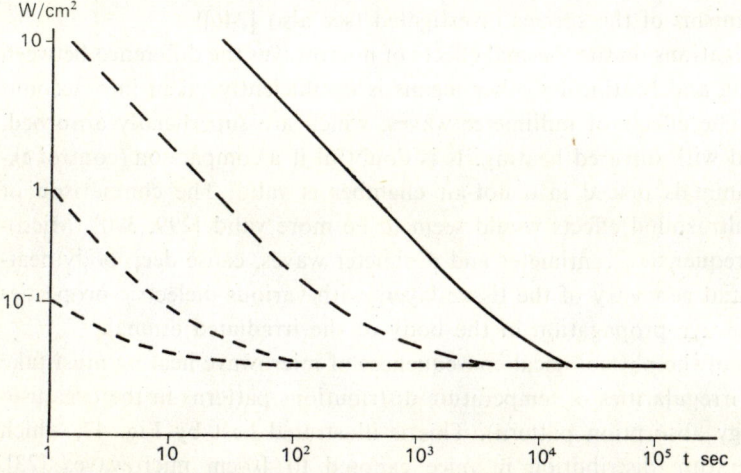

Fig. 46. Time needed for temperature increases by 5°C in various animals (dashed line and finely dashed line) and man (continuous line) expected at various power densities. Approximate curves drawn from data in [232].

Using this approach, it may be shown that the power density, at which infinite exposure time is possible, is similar for all animal species and amounts to about 30 mW/cm². If the denominator of formula 31 is equal to 0, the power density for infinite exposure time I_∞ is

[33]
$$J_\infty \simeq [S\alpha(\varphi + \Delta T) - M]\frac{3J}{S} \text{ W/m}^2$$

The basal metabolic rate correlates with body surface, and the metabolic heat may be approximated as

[34]
$$M = 2M_B S$$

where M_B is the metabolic rate per square meter of body surface. Expression 33 for infinite exposure time becomes

[35]
$$I_\infty \simeq [\alpha(\varphi + \Delta T) - 2M_B]3J \text{ W/m}^2$$

Therefore, the power density for infinite exposure time depends mainly on the metabolic rate per square meter of body surface. This value shows slight variations for most mammals.

A comparison of expression 35 and the curves in Fig. 46 with empirical results demonstrates that the calculated values are lower than the empirical ones for larger animals, both for dogs and man. Man is characterized by the high efficiency of his thermoregulatory mechanisms [217, 258]. Calculated values are distinctly higher than empirical ones for mice [23, 25]. These data may be used for preliminary planning of experiments on thermal effects of microwaves and microwave effects not accompanied by body-temperature increases. The difference between the calculated and empirical values may also be used for assessing the efficiency of the thermoregulatory mechanisms of the species investigated (see also [246]).

In many investigations on the thermal effects of microwaves the difference between microwave heating and heating by other means is insufficiently taken into account [102, 257, 258]. The effects of millimeter waves, which are superficially absorbed, may be compared with infrared heating. It is doubtful if a comparison (control experiments) with animals placed in a hot-air chamber is valid. The comparison of microwave and ultrasound effects would seem to be more valid [299, 390]. Microwaves of lower frequencies, centimeter and decimeter waves, cause deep body heating, with the spatial geometry of the tissue layers with various dielectric properties influencing microwave propagation in the body of the irradiated animal.

Considerations of the physiological consequences of microwave heating must take into account the irregularities of temperature distributions patterns in the organism (microwave energy absorption pattern). This is illustrated best by Fig. 47, which represents temperature distribution in mice exposed to 10-cm microwaves [23]. The drawing serves only as an example, because temperature distribution depends on the interplay of two factors — the relationship between the wavelength and the

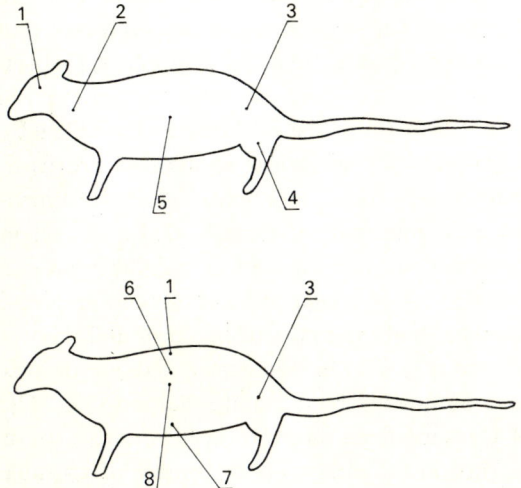

Fig. 47. Distribution of temperatures in mice exposed to "acute" 3-cm irradiation. Top: 1, 45.4°C under the skin; 2, 42.6°C in the esophagus; 3, 41.7°C rectal temperature; 4, 41.8°C femoral muscle; 5, 43.6°C liver. Bottom: 1, 42.9°C dorsal muscle; 6, 46.3°C dorsal subcutaneous tissue; 3, 40.3°C rectal temperature; 7, 40.8°C abdominal cavity; 8, 42.1°C kidney. According to [23] and [25], redrawn after [21].

spatial geometry of tissue layers (skin, subcutaneous tissue, muscles, and internal organs; see Chapter 3) and the blood supply of particular layers or organs. The first of these two factors must be considered from the point of view of microwave propagation and absorption. The second factor depends on both the anatomical structure, i.e., the vascularization of the structure considered, and on the vascular reaction to microwave irradiation. Several investigations pertaining to these problems have been carried out; the interpretation of the results should be restricted, however, to the situations investigated. Numerous comparative investigations on temperature distributions in various human and animal tissues during irradiation at various wavelengths can be cited [48, 91, 154, 157, 187, 188, 226, 240, 286, 300-303, 378, 393, 394, 407, 410, 411, 424, 448, 489, 499, 500]. Differences between the effects of whole-body irradiation and irradiation of segments or parts of the body should be kept in mind. In both situations secondary effects within the whole system (body) should be considered. Local temperature increases may influence the function of nonirradiated parts (nervous reflexes, influence of metabolites).

It seems superfluous to cite individual results and, moreover, space does not permit it. In view of this, an attempt will be made to present a generalization of the characteristic temperature effects of microwave radiation, stressing those aspects that seem to be of prime importance in the evaluation of biological consequences-secondary effects.

In particular layers of the body, temperature gradients may occur because of differences of depth of penetration and energy absorption, depending on the wavelength of the incident radiation. The temperature of the skin and subcutaneous tissue is influenced by the temperature of the air. The temperature of subcutaneous tissue may be much higher than that of the skin surface because of air cooling. The survival time of rats exposed to high power densities was markedly prolonged when the animals were placed in chambers flushed continually with liquid-nitrogen--cooled air [470]. Theoretical considerations and model experiments indicate that heat conduction in subcutaneous tissue corresponds to 0.005 kcal/cm·s [91], increasing linearly with exposure time. The waves may be transformed, however, in fatty tissue [225], and depending on the thickness of the fat layer and wavelength, standing waves may be created. When physiotherapeutical applicators are used, excessive heating of subcutaneous tissue in fatty tissue may occur, when the dimensions of the aperture source are less than one half-wavelength [211]. In certain irradiation conditions excessive temperature may be induced in the subcutaneous tissue [393, 394] or in the skin [393, 499]. This may lead to local thermal necroses.

The temperature distribution in superfical body layers is of practical and theoretical importance. Heat perception and pain reactions in man exposed to microwaves are different from the reactions induced by heating by infrared radiation [92, 392], which is easily understood in view of the superficial localization of sensitive nerve endings. The practical importance lies then in the alteration or absence of warning sensations. From the theoretical point of view, thermal stimulation of nervous receptors at various depths may influence the course of the physiological generalized reaction to microwave exposure [160, 284].

The distribution of energy absorption and the resulting temperatures in deeper layers muscles and bone tissue have been investigated mainly from the point of view of physiotherapeutical applications [472]. Newer investigations concern mainly the characteristics of the source applicator influencing the resulting relative temperature distribution pattern [295, 296, 300, 302]. Theoretical considerations and the studies on phantom models equivalent to skin, subcutaneous tissue, muscles, and bone of microwave energy absorption depending on the characteristics of the source and irradiation conditions, carried out by *Guy et al.* [210, 211], merit special attention. Such studies may be used to obtain more precise data on the heat-balance characteristics of microwave-irradiated animals and to correct the approximate solutions cited previously, especially in respect to superficial versus deep body heating [232, 246]. As space does not permit elaboration this point, the reader is referred to the original papers by *Guy* [210, 211], which contain pertinent references.

A complete evaluation of temperature effects in muscles and internal organs demands the quantitation of cooling by blood flow. The investigations on local [177, 196, 257, 451, 456, 576] and generalized [93, 96, 428, 429] circulatory reactions to microwave exposure are far from complete. Examination of the role of circulatory reactions in muscles may be difficult in view of the report that both vasodilatation and hyperemia, or vasoconstriction and ischemia, or even absence of changes in blood flow may occur during microwave irradiation [196]. This single report based on $Na^{131}I$ absorption in man during local irradiation with 12.4-cm waves indicates the need for further studies. Plethysmographic studies indicate that usually a summary increase in blood flow by about one third occurs [451, 456].

The presence of focal lesions following whole-body irradiation, mainly in the brain, liver, and kidneys, has been demonstrated by many authors [191, 234, 249, 309, 326, 379, 407, 409, 554, 557, 600, 601]. Limited foci of necrosis, hyperemia, or stasis and degenerative changes were reported. These observations may be explained by local overheating (high temperature increases). This leads to the assumption that in the species examined (mice, rats, guinea pigs, rabbits, and monkeys) local amplification (standing waves, interference) may occur at the wavelengths used, i.e., from the 3- to the 12-cm waveband. It should also be pointed out that metallic inclusions in the body not only distort the field, but also may cause burns in surrounding tissues because of overheating of the metal [14, 396].

Focal lesions in the brain, indicative of local excessive temperature increases, merit special attention [409, 554, 557]. Irregular field distribution patterns within the cranium and brain may be deduced from such observations [12, 485]. A detailed analysis of this problem was carried out by *Shapiro et al.* [493]. These authors analyzed the distribution of 3000-MHz microwave fields in a multilayered lossy sphere based on the dimensions of cranial structures in macacus monkeys. The following layers were considered: brain tissue, cerebrospinal fluid, dura, bone, fatty subcutaneous tissue, and skin. Solutions for the internal and scattered field distributions were presented in a form amenable to machine computation. These solutions could be applied and developed further for the analysis of temperature distribution in multilayered objects with curved surfaces, i.e., laboratory animals irradiated at various wavelengths. The authors consider also human cranial structure. The main conclusion which may be drawn from this work is that the semi-infinite-slab model of *Schwan* and *Li* [480, 481]; (see also Chapter 3) is inadequate for such objects, in which the relation between the curvature radii and the wavelength remains within the limits of 0.05 to 5. The authors illustrate the biological significance of their findings by the comparison of basal gray-matter metabolic heat about 3 cal/g·s with the heat produced by 2450-MHz microwave energy absorption at an incident external power density of 10 mW/cm^2. Mean absorption in the human head corresponds to 0.25 cal/g·s and 0.6 cal/g·s in the monkey. Local absorption peak value corresponds in man to 1 cal/g·s and 3 cal/g·s in the monkey. Figures 35, 36, and 37 (compare the figures and text on pp. 57-60) illustrate the mean absorbed power density in the monkey cranium according to *Shapiro et al.* [493].

Finally, it should be pointed out that realistic models and quantitative solutions of the complex relation between the incident power density (as measured in air) and the resulting temperature effect in biological objects have been elaborated only in recent years. It seems that the solutions proposed are one of the necessary successive steps in the approximation of the real situation. The solutions cited at the end of this section seem to be very promising. Machine computation seems to be unavoidable for quantitation of temperature effects.

Effects on the Nervous System

The majority of investigations on the influence of microwaves on the bioelectric function of the brain, conditioned reflexes, and morphology of the nervous tissue have been carried out by Soviet authors. A review of the results obtained is presented by *Cholodov* [72], who is quoted extensively by *Presman* [438], whose monograph is easily accessible to the English-speaking reader. Although we wish to avoid a repetition of the data contained in *Presman's* monographs (see also *Petrov* [425]), a general brief survey of the Soviet investigations on this subject must be presented.

Cholodov [37, 72, 73] investigated the bioelectric brain function of animals subjected to electromagnetic and magnetic fields in various conditions. EEG registration demonstrated an increase of slow, high-amplitude waves and the occurrence of spindles following exposure to PW or CW 12- and 52-cm microwaves at high power densities 100 to 1000 mW/cm^2. Preconvulsive discharges were also observed. Lethal effects were obtained earlier by exposure to PW. The excitability of the visual cortex region increased at power densities of 100 to 300 mW/cm^2, and decreased at 1000 mW/cm^2. In 31 percent of the animals examined desynchronization of the bioelectric activity of the cortex was observed at the moment of switching the transmitter on or off, which the author explains as a reaction to microwave stimulus. When the brain was cut through at the midbrain level, the EEG of irradiated animals showed a tendency to normalization. Simultaneous caffeine administration and microwave exposure increased the frequency of convulsions. Registration of neuronal activity following irradiation indicates a decrease in the number of active neurons. *Cholodov* [73] examined the EEG of animals with severed optical nerves, with the brain trunk severed at the midbrain level, and with destroyed visual or auditory cortex which were subsequently repeatedly irradiated 15 to 20 times for 3 min. Similar changes in bioelectric activity were observed on irradiation in all these experiments; however, the mean latency time was more prolonged in animals in which the auditory cortex was destroyed.

Cholodov [72] claims that microwaves influence specifically different brain structures. The high exposure levels used make it probable that temperature effects influenced the results. Simultaneous irradiation and EEG recording make artifacts caused by field interaction with electrodes highly probable. The reaction on switching the transmitter on or off may serve as a suspect example. Thermal stimulation by selectively heated electrodes and the field distortion caused by their presence make the evaluation of the findings difficult. Because these objections were realized and raised early by *Subbota* [514], many Soviet authors tried to verify the described observations in more critical experiments or by using other methodological approaches to the examination of microwave effects in the nervous system (see *Petrov* [425]).

Regrettably it is not always possible to ascertain from the published texts if sufficient precautions were taken. Reviews of the data obtained show a certain consistency of the findings obtained in different experimental situations (see *Presman* [438 and *Petrov* [425]). A short additional review of electrophysiologic data is necessary before any attempt at the evaluation of their significance is made.

Knepton and *Beischer* [271] reported an increase of frequency and amplitudes in the EEG of chronically irradiated animals. *Bytchkov* and *Syngajevskaja* [59] irradiated 160 rabbits, simulating the conditions occurring during circular radar observation or at a fixed azimuth (sector), and submitting the animals to continuous stationary field or intermittent (nonstationary field) exposure. Only slight deviations from normal tracings were reported; no differences were demonstrable between the irradiated groups. *Bytchkov* [57] examined the EEG, chronaxiometric findings, and reflectory reactions in microwave-irradiated rabbits, cats, mice, and frogs. Changes in EEG recordings and abnormal reflexes were seen. The animals became more sensitive to strychnine administration. According to this author, microwave exposure affects polarization phenomena, the thermal effects favoring depolarization. Temperature effects may explain the above cited findings.

The diverse findings on the bioelectric function of the brain in microwave-irradiated animals may be explained by the diversity of exposure conditions and species specificity of reactions, influenced also by the size and anatomical structure of the animals used. At exposures of 20 mW/cm^2 no generalized body temperature rises were observed, which does not preclude local thermal effects. *Gvozdikova et al.* [213] reported different EEG findings in rabbits exposed in identical conditions. In 19 animals a slowing down of bioelectric rhythms and an increase in amplitudes were observed; in 21 animals, a frequency increase and amplitude decrease; and in 23 animals no changes were observed (12.5-cm microwaves at power densities of 0.02 to 50.0 mW/cm^2). Similar slowing down and amplitude decrease and frequency increase with amplitude increase were observed in the EEG of animals irradiated with 52-cm microwaves.

Bytchkov and *Syngajevskaja* [60] found that ganglioplegic drugs counteract microwave irradiation effects, preventing convulsions after administration of low strychnine doses. These authors maintain that microwave irradiation exerts an inhibitory influence on cholinergic synapses and causes a decrease in cholinesterase activity in the brain. Similar findings were reported by *Nikogosjan* [404], the cholinesterase activity depending on the duration of the experiment repeated long-term exposure. Low power densities do not affect cholinesterase activity.

Visual and frontal cortex acitivity was examined by *Bavro* and *Cholodov* [37] in animals exposed to 12-cm microwaves at 1 and 300 mW/cm^2. Slow, high-amplitude potentials appeared in visual cortex activity, and high-amplitude discharges in the frontal cortex. Changes appeared during the few minutes following the switching off of the transmitter. The activity of the visual cortex was higher at lower power density and decreased at high power density. Following caffeine administration and microwave irradiation, convulsions could be induced; microwave irradiation alone did not lead to the appearance of convulsions. Mechanical stimulation of the cortex and microwave irradiation caused convulsive discharges in 80 percent of animals; light stimulation and irradiation caused convulsions in 60 percent. In principle, 1 mW/cm^2 may be considered as an athermal dose, with 300 mW/cm^2 being decidedly thermogenic and leading to temperature increase. Differences in the effects observed at both power density levels and changes in the reactions of irradiated animals to various stimuli and/or drug administration merit further investigations. The role of specific i.e., nonthermal mechanisms in such phenomena has been suggested by certain authors.

Gordon [184, 190] is of the opinion that changes in EEG tracings of irradiated animals should not be looked upon as specific. Amplitude and frequency changes appear during the first few minutes of irradiation. The effects obtained are more pronounced as the wavelength increases. Long-term irradiation at low (athermal) power densities induces reactivity changes expressed by a lowering of the threshold for convulsions provoked by various stimuli. *Gordon* explains the appearance of theta waves by effects in subcortical structures, specially the diencephalon.

The results of these and other investigations of Soviet authors [1, 37, 59, 71-73, 195, 247, 291, 314, 315] on microwave effects on the EEG of rabbits may be summarized as follows [438]:

1. After a certain latency time of a few tens or hundreds of seconds, numerous small, high-amplitude waves appear during microwave irradiation.

2. Desynchronization of long-term duration appears, basic amplitudes decrease, and the frequencies increase.

3. Switching the microwave source on and off causes a short-duration desynchronization as described in point 2 (artifact?).

4. After termination of irradiation, after a varying time interval, signs of desynchronization appear.

5. Preconvulsive spikes of high frequency and amplitude may be evoked.

6. The basic changes are similar and of the same general character at various power densities, beginning at 2 mW/cm^2 and ending at 1000 mW/cm^2, in the whole microwave band.

7. Irradiation of various body parts (except the head) and lateral irradiation cause a decrease in biopotentials and the appearance of high-frequency rhythm, i.e., desynchronization of the bioelectric brain function.

8. The character of the observed changes is not influenced by experimental injuries to the hypothalamus, thalamus, and reticular formation or by destruction of the visual, auditory, and olfactory senses.

9. Adrenaline acts synergistically to microwave irradiation.

10. Microwave irradiation increases the reaction of rabbits to light stimuli.

As stated previously the presence of implanted electrodes during irradiation and simultaneous EEG recording and irradiation make it difficult to evaluate how far these findings may be influenced by artifacts and/or thermal stimulation of various brain and body regions by heated electrodes. Electrical stimulation also cannot be excluded. Critical reevaluation by appropriate control experiments is necessary.

Interesting experiments were described by *Bessonova* [43]. The excitability of the sciatic frog nerve changes during irradiation at 2 mW/cm^2, i.e., at an admittedly nonthermal level. Oscillographic investigations on the function of ganglia and extraganglionic neurons in microwave-irradiated animals were carried out by *Gavrilova* [174]; an increase in lability and a decrease in the amplitudes of the bioelectric potentials were noted.

The influence of local head irradiation 2 to 10 min, 320 MHz in the monkey was investigated by *Bach et al.* [12]. The longest survival time was 10 min, 55 s. The EEG recordings may be interpreted as characteristic of temperature increase effects caused by irradiation. A period of inactivity (immobility and lack of reaction to light and pain stimuli) was followed by clonic and tonic convulsions, which were succeeded by a terminal period of prostration. Similar effects were obtained by *Austin* and *Horwath* [9] in dogs after 2450-MHz head irradiation. Tonic and clonic convulsions preceded death; the survival time was dependent on power density. Irradiation at the level of the IV-XII thoracic vertebrae led also to death, but without preceding convulsions. These authors explained their findings by thermal (temperature) effects, although not excluding the existence of other effects.

The neurologic effects of 3-cm microwave irradiation were investigated in decorticated or anesthetized cats by *McAfee et al.* [340, 347]. Spinal reflexes were unimpaired in the decorticated animals. A small segment of the sciatic or radial nerve was prepared and heated with microwaves or infrared radiation up to 45°C. Similar physiological responses were observed in both cases-blood pressure increase, heating of surrounding tissues. Cooling the irradiated nerve demonstrated the temperature dependence of the induced effects. Investigations on microwave irradiation of skin sensory nerves led to the same conclusions: 3- and 12-cm microwave irradiation evoke only temperature-dependent responses [343, 347]; heating by other means leads to identical effects. No signs of high-voltage, spontaneous activity of myelinate fibers were observed. Thermal microwave effects and local nerve heating by any other means induce vegetative responses of the same character. Granted that the investigations of *McAfee et al.* demonstrate the temperature-dependent effects induced by microwave heating, these observations are insufficient to exclude other, nonthermal mechanisms which may be operative. Long-term, low-dose, irradiation--induced effects cannot be explained on the basis of these experiments.

It should be pointed out that the experiments cited arose out the observation on the analeptic effects of 10,000-MHz microwave radiation applied to the head of anesthetized rats, rabbits, cats, and dogs [340, 342]. Primarily, this effect was explained by direct action on deep brain structures, the reticular system mainly, and was nonthermal in nature. Simultaneously, a rapid drop in blood pressure was observed in rats and rabbits and a blood pressure increase in cats. Similar effects may be obtained, however, by local irradiation of peripheral parts of the extremities

[160, 344], which precludes any direct action on the central nervous system. In this manner it could be demonstrated that these effects are mediated by peripheral nerve stimulation. Xylocaine administration, which blocks nerve conduction, abolishes analeptic microwave effects [342]. Control experiments with conduction, convection, or infrared peripheral nerve heating up to 45°C demonstrate the temperature dependence of the effects. It may be concluded that the heating of peripheral nerves to

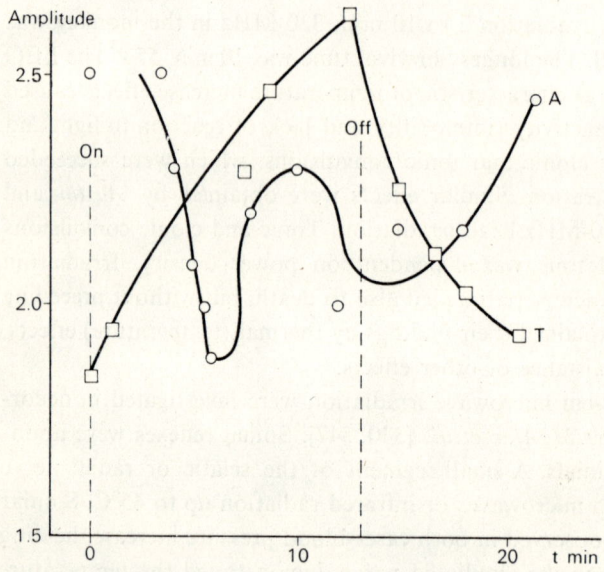

Fig. 48. Changes in biopotentials in the brain of a cat (A) and in temperature of the heated sciatic nerve (T). According to [160].

this temperature induces a central nervous system mediated secondary response to microwave irradiation. According to *McAfee* [343], this effect depends on fiber heating and not on nerve-ending stimulation. A similar mechanism could be responsible for the EEG changes induced in cats by sciatic nerve irradiation with 10,000 MHz microwaves observed by this group of authors [160], as shown in Fig. 48. Similar mechanisms may be responsible for reactions induced by the irradiation of peripheral body segments in rats [314, 315].

Another approach lies in the numerous investigations on conditioned reflexes in 10-cm-microwave-irradiated animals. *Svetlova* [527] reported that 1-h irradiation at 0.2 mW/cm² caused a decrease in reflex-conditioned saliva secretion in dogs on the irradiated side. *Minecki et al.* [382, 383] demonstrated changes in conditioned--reflex responses in 10-cm-microwave-irradiated rats. These authors used a CW generator with a horn antenna, irradiation conditions corresponding to the far-field zone. Changes in conditioned-reflex responses were demonstrated following only a single exposure; repeated exposures led to sings of excitation and, later, of inhibition (in the Pavlovian sense). Disturbances in conditioned reflectory function were more or less pronounced, depending on power density 16 to 94 mW/cm^2. *Lobanowa* and *Tolgskaja* [321] observed conditioned reflexes in rats exposed to 10-cm microwaves at 10 mW/cm^2 for 45 days from 30 min to 1 h daily. Following a few exposures, excitation and, later, inhibition (in the Pavlovian sense) were observed. During the final period of the experiment, frequent absence of reaction (motoric) to the conditioned stimulus (light, sound) was observed. Slight individual differences among animals in the time course of reactions were observed. It is interesting to note that, in cats irradiated in analogous conditions, changes in the histological picture of the motor regions in the brain cortex were observed by the same authors. In histological sections impregnated with silver, according to the Golgi method, a decrease in synaptic connections (number of nerve endings on the neuron body) and changes in the appearance of dendrites were seen. Thickenings of dendrites, first far from and, later near the neuron body were seen. A finding of importance is the reversibility both of functional and morphologic changes; they disappeared between the tenth and twentieth day after termination of irradiation.

The functional disturbances in conditioned-reflex responses in irradiated animals as described by *Minecki* [382, 383] and *Lobanova* and *Tolgskaja* [321] are well documented. No objections, similar to those raised in the case of EEG registration, can be made on the ground that the experimental biologic technique influences irradiation conditions. As regards the morphologic observations of *Lobanova* and *Tolgskaja* [321, 556], it should be pointed out that the Golgi technique of silver impregnation may be misleading; frequently, histologic-technique artifacts are met with.

Morphologic changes do occur in the brains of irradiated animals. Their severity depends on both power density and wavelength; in many cases the morphologic pictures are indicative of focal thermal lesions. A detailed description is given by *Tolgskaja* and *Gordon* [554]. A brief description will be given, based mainly on the observations of *Tolgskaja, Gordon*, and *Lobanova* [556]. In rats repeatedly irradiated (75 days) at high power densities (40 to 100 mW/cm^2), brain hyperemia, pycnosis, and vacuolization of nerve cells were observed. A proliferative reaction of microglial cells was noted. Similar, but less pronounced, changes were seen following exposures at 10 to 20 mW/cm^2. According to the cited authors, no signs of hyperthermic (temperature) lesions or microglial reactions were noted. Comparison of 3- and 10-cm microwave effects at the same power density of 10 mW/cm^2 has shown that the changes induced are similar in both cases, but less pronounced following 3-cm microwave exposure. Similarly, PW irradiation caused more distinct changes in the morphologic picture of brain histological preparations than CW irradiation, the overall character being in both cases the same. Several days after termination of the irradiation period no changes were demonstrable, which points to the full reversibility of morphologic signs of injury caused by such irradiations. The authors stress that the changes described lack specificity; similar phenomena may be observed following the administration of various poisons [555].

Piervuchin and *Triumfov* (according to [444]) described degenerative lesions in nerve cells in the brain of microwave-irradiated rabbits. Brain ventricles and blood vessels were distended; extravasations of blood were noted. *Oldendorf* [409] made similar observations, and described lesions of both gray and white matter in the brain of rabbits exposed to high-power-density, 12-cm microwaves.

Minecki and *Bilski* [379] described degenerative changes in the brain of mice irradiated with 10-cm CW microwaves at various power densities 16.5, 30, and 54 mW/cm^2. Similar pictures accompanied by signs of inflammatory reactions were reported by *Niepolomski* and *Smigla* [403], *Pitenin* [431], and *Horai* [234]. Changes in the morphology of nerve endings thickening or segmentation of the myelin sheath in various internal organs, muscles, and skin have been described [557]. *Cholodov* [72] and *Sokolov* (according to [21]) point out proliferative glial reactions. *Orbelli* (according to [440]) draws attention to the proliferative reactions of small blood vessels in the thalamic and hypothalamic regions.

Cholodov [72] reports glial reactions in rabbits after only 1-h exposure to high--intensity magnetostatic fields. After 12 h edema of processes of the glial cells and perivascular changes became apparent. Dystrophic degenerative changes in glial cells followed after 72 h. According to *Cholodov*, these changes may be responsible for signs of inhibition.

The preceding experiments demonstrate a relatively slight effect for 3-cm microwaves, as opposed to marked changes in the nervous system induced by exposure to 10-cm waves. The intensity of the induced changes depends on the power-density duration of exposure and modulation. Ten-centimeter microwaves may affect the bioelectric function of the brain [58, 185, 346, 347] and disturb the normal relationship between the phenomena of excitation and inhibition (in the Pavlovian sense) of the higher nervous function [234, 378, 510, 512, 514, 516]. This last effect may also be obtained by exposures to low power densities, which may indicate that nonthermal (temperature-increase independent) mechanisms of microwave interference with the nervous system function are involved. Certain authors feel that thermal temperature increase effects may be responsible [9, 253, 342]. Morphologic alterations, as described previously were demonstrable in situations where such power densities were used, at which temperature increases are to be expected [1, 72, 379, 554, 557]. Nevertheless, participation of additional nonthermal mechanisms cannot be excluded on the available evidence.

The investigations on unconditioned and conditioned reflexes in microwave-irradiated dogs carried out by *Subbota* [512, 513] and *Svetlova* [525, 527] deserve special mention. Classic Pavlovian methods were used. The authors determined the time of initation of saliva secretion following the conditioned stimulus latency time and the number of drops secreted. Following lateral irradiation with 10-cm waves for 2 h at a power density of 1 mW/cm^2, or within the 1- to 5- mW/cm^2 range, the intensity of the response increased on the opposite side, and the latency time was shortened. Differentiation of stimuli remained unaffected. On subsequenty exposures, sings of stimulation became more apparent, and paradoxical reflexes appeared. Prolongation of the experiment led to a tendency to normalization. Following a total of 70 h of exposure 35 days, 2 h daily, conditioned-reflex responses became identical as before irradiations. According to *Subbota* [514], this indicates the gradual adaptation of dogs to successive microwave exposures. This phenomenon has a time course analogous to the temperature responses described previously.

Special importance is attached by the cited authors to the asymmetry of conditioned- and unconditioned-reflex responses induced by lateral microwave exposures [516]. Immediately following termination of irradiation in the 10-cm waveband, disturbances became apparent on the side opposite to that which was irradiated. Following irradiation with decimeter waves (precise wavelength not specified by the authors) at 1 mW/cm^2, disturbances were noted on the irradiated (exposed) side. Decimeter--wave exposures led more frequently to signs of neurosis, as evidenced by disturbances of equilibrium between the processes of excitation and inhibition (in the Pavlovian sense) in the nervous system.

The cited authors suppose that interaction with peripheral nerves is responsible for microwave-induced effects on conditioned-reflex responses. The main reason is that on lateral irradiation the more superficially absorbed 10-cm waves influence the response on the side opposite to that which was irradiated, and the more penetrating decimeter waves influence the irradiated side, which may indicate a direct interaction with brain structures. In the case of decimeter waves, interaction with peripheral nerves may be demonstrated by irradiation of the trunk, with the head screened. It is interesting to note that more distinct disturbances were observed in animals with the head screened. In view of the rather capricious distribution of fields within cranial structures see p. 59 and the lack of experimental verification of the efficacy of head screening (reflections or scattering within the body causing microwave energy absorption in the head) the preceding explanation needs not necessarily be true.

According to *Subbota* [514], irradiation of peripheral parts of the body may influence the function of the central nervous system (conditioned reflexes), or through interaction with peripheral nerves or by induction of other changes in irradiated organs or tissues, influence indirectly the central nervous system function, for example through changes in metabolism. This concept would agree well with the notion of "secondary microwave effects."

Meter waves at 1-mW/cm^2 power densities induced symmetrical changes in conditioned-reflex responses in dogs.

Changes in ambient air temperature may also influence conditioned reflexes. Nevertheless, it is possible to obtain a fixed level of response. *Subbota* and *Svetlova* [517] held dogs with a conditioned-reflex response in a hot room at 35°C and obtained by repetition of the unconditioned and conditioned stimuli a level of response equal to that of the preceding period (before exposure to the environment). A single exposure to 10-cm microwaves at 1 mW/cm² caused a decrease in response similar to that induced by the environment. Noise may be used as an interfering factor in conditioned-reflex experiments. Single microwave exposures caused changes in response to noise in adapted dogs. These experiments led to the concept of the "desadaptive effects" of microwave irradiation. These effects are demonstrable for 24 to 48 h following single exposure.

Three-centimeter microwave head or whole-body irradiation at power densities of 1 to 10 mW/cm² does not interfere with conditioned reflexes in dogs; demonstrable effects were obtained at 20 to 30 mW/cm².

Simultaneous exposure to decimeter and meter waves at 0.5- or 1-mW/cm² power densities for 30 min or 1 h daily induced effects similar to, but more pronounced than, decimeter-wave exposure alone. Summing up these results *Subbota* [514] points out that decimeter waves (wavelength not specified) induced pronounced changes in conditioned-reflex response and signs of "experimental neurosis" (in the Pavlovian sense) at power densities of 0.5 to 5 mW/cm².

Corroborative evidence may be found in *Kicovskaja's* experiments [264, 265]. Rats were irradiated with 10-cm, 3-cm, and millimeter microwaves at 10 mW/cm², and their reaction to an auditory stimulus (bell sounding) was examined. The animals reacted to the bell with motoric excitation before irradiation. The response occurred 5 to 15 s following the stimulus. Movements were registered using a Marey's drum placed under the cage floor and a kymograph. The excitation decreased gradually and disappeared after 10 to 20 s. Following this, a second phase of excitation occurred. According to *Kicovskaja*, the auditory stimulus causes an excitation on higher central nervous system levels. Prolonged action of the stimulus causes inhibition, which induces, on receding, the second excitation phase. Depending on the intensity of the inhibition processes in individual animals, monophasic or biphasic reactions are observed. Twenty-five to thirty daily microwave exposures, each of 1-h duration, induce an increase in the intensity of inhibition phenomena, as demonstrated by an absence of response in animals with an anterior biphasic reaction or by substitution of a biphasic reaction for the monophasic. This effect was consistently obtained by irradiation with 10-cm microwaves, but only in some of the 3-cm-irradiated animals, and could not be induced with millimeter-wave exposures.

Similar results to those reported by *Kicovskaja* [264, 265] were obtained at the University of Kansas [256, 269, 270]. Rats were placed in a Tappan R-3-L microwave oven (12.25 cm), and irradiated daily for 124 days. During 1-hr irradiation sequences of 1-min periods of power emission (50 percent) were alternated with 1-min periods without power emission (50 percent); i.e., intermittent exposure was applied. The power densities were calculated from calorimetric determinations of energy absorbed in phantom models made from plastic bags filled with physiological saline solutions. The authors introduced a dose concept based on the absorbed energy per body weight and body surface exposed to microwaves (for details, see the original paper). Using a more accepted criterion of exposure level, the conditions could be approximated to exposures at power density levels of 2.5, 5, 10 and 15 mW/cm^2. The natural behavior of rats was used to investigate the effects of irradiation. The tongue-licking reflex was reinforced by placing a drop of sugar water on the snout. A tonal stimulus of 525 Hz was combined with sugar water or the drop was placed at the moment of sound cessation. Licks became more frequent with increasing power densities, and on further increase become less frequent. The conditioned response disappeared at higher power densities. In animals with a stable response to the 525-Hz tone, the combination of this stimulus with microwave irradiation caused a drop in responses to 68 percent of control values, in spite of retaining the full ability for signal discrimination. At higher power densities the animals became immobile and lay down. Signs of prostration and of what was called by the authors "flaccid paralysis" were observed. It should be stressed that the expressions "prostration" and "flaccid paralysis" are used by the authors in a purely descriptive sense without any pathogenetical implications. These signs disappeared completely a few minutes after termination of irradiation. At high power densities, rectal temperature increased to more than 41°C (end thermometer scale). The authors point out that the relationship between the wavelength and object size favored energy absorption. Histological examinations of the brain on termination of the experiments did not reveal any lesions.

Motor activity, excitability, and susceptibility to convulsions provoked by tonal signals or electric shock in microwave-irradiated rats were examined by *Korbel* [280, 282]. The animals were subjected to irradiation with waves of variable frequency (changing within the range from 300 to 920 MHz and back) at a power density of about 0.76 mW/cm^2 during 24 h (examination periods excepted) for 47 days. In comparison with control animals, motor excitation was observed during the initial periods of the experiment up to the fifteenth day. A decrease in activity became evident later, beginning on the twenty-first day and becoming markedly pronounced between the thirtieth and forthieth day. The animals became emotive, as evidenced by reaction to bright light. The latency period and the clonic phase of the convulsions induced by a 0.5-s electric shock (60 mA) were distinctly prolonged. No differences in the convulsions provoked by a tone of 140 dB during 3 min were noted between irradiated and control animals. Weight and water intake were also not influenced. Learning facility (water maze) was decreased in the experimental group; a longer time was required and more errors were made. The weight of the adrenals was significantly decreased in comparison with controls. Exposure to 0.5 mW/cm^2 for 40 days caused only a decrease in motor activity. In a further series of experiments involving exposures to low power densities from 0.43 to 0.15 mW/cm^2 in two frequency ranges (320 to 450 MHz and 770 to 900 MHz), it was found that the lower--frequency-range exposures affected the activity of rats more than higher-frequency exposures. Examinations made on successive days of the experiment indicate that a cetrain duration of the experiment is needed to observe the described effects.

In the present author's opinion, this means that cumulative effects of low-dose exposure are conceivable. In a further set of experiments the same group of authors [270] demonstrated in similar exposure conditions Tappan R-3-L microwave oven, 12.25-cm waves that rats may be conditioned to recognize as a warning cue 1-min exposures at 6.4, 4.8, 2.4, and even 1.2 mW/g, which correspond approximately to 20, 15, 7.5, and 3.75 mW/cm^2 for calculation of the dose, see [256]. No measurable increases in whole-body i.e., deep rectal temperatures were detected by the authors, whose conclusion is that "whatever the mechanism of detection ... mammals are sensitive to something that inheres in or accompanies illumination by microwaves at low levels of available power."

Hearn [220] found changes appearing after several exposures in the discrimination of flicking light in rate repeatedly irradiated at frequencies of 300 to 920 MHz.

The preceding findings of American investigators agree with the results reported by Soviet authors, at least to the extent that microwave irradiation may cause disturbances in the function of the central nervous system.

In the experiments of *Justesen* and *King* [256] no positive operant reaction to microwave exposure was obtained. *Frey* [164, 165] obtained conditional reflexes to microwave irradiation in cats. One possible mechanism involved could be the hearing of radio frequencies and a reaction to microwaves as an auditory stimulus. The two papers cited present a review of *Frey's* investigations and represent the decided standpoint taken by this author on the interaction of microwaves with the nervous system. The reader is encouraged to read these publications as well as the references cited [162, 165].

Frey developed special animal restraints, new types of electrodes, and an original system for registration of the bioelectric function of nervous tissue during irradiation [169, 171]. The discussion of the requirements for the validity of such an experiment as well as the data that should be included into its description to make evaluation and reproducibility possible may be usefully consulted [163]. Among the results reported, the possibility of obtaining evoked brain potentials in cats by microwave irradiation [162] merits special attention. The power density threshold values for this phenomenon are reported as 0.03-mW/cm^2 mean power density and about 60-mW/cm^2 peak value. *Frey* points out the importance of considering the peak values during irradiation with pulsed microwaves, maintaining that they play an important role in determining the responses of the nervous system. Screening of the head or of the body trunk during irradiation was used to demonstrate that microwave-evoked brain potentials are obtainable only by direct irradiation of the head. Frequencies in the range from 1.2 to 1.525 GHz were used; a more pronounced effect was obtained at lower frequencies. Changes in wave polarization did not influence the results, which depended, however, upon the pulse repetition rate. Up to 50 Hz a distinct series of evoked potentials was obtained; at higher repetition rates a series of discharges was obtained. Differences occurred among the tracings obtained from various brain structures; evident signs of the influence of microwave irradiation were seen in the function of caudal parts of the reticular structure. These results seem to indicate that direct interaction of microwaves with various brain structures may occur. *Gavrilova's* [174] investigations point also to the possibility of influencing the function of peripheral pre- and postganglionic neurons by low--power density exposures. A decrease in amplitudes and lability of the bioelectric function were noted.

Further evidence of changes induced in the brain by microwave irradiation may be found in a series of experiments carried out by *Barański* and *Edelwejn* [19, 20, 30, 142, 144]. The influence of single and repeated exposures to 10- and 3-cm microwaves on the bioelectric function of the brain (EEG), its histological picture, and the activity of brain enzymes were examined in rabbits and guinea pigs. Repeated exposures were made at power densities of 5 to 7 mW/cm^2, and single exposures at increasing power densities, beginning at 5 and ending at 30 mW/cm^2. Temperature measurements were made at termination of the irradiation period (1 to 2 h) under the skin and behind the frontal bone, as well as at various points of the brain in a special series of "temperature control animals." Significant temperature increases (thermocouples, 0.2°C sensitivity) were noted after exposures to 20 mW/cm^2. The animals were irradiated in far-field conditions in an anechoic chamber in plexiglas (methacrylate) cages, placed with their heads toward the source (horn antenna).

Single exposures at an increasing power density of pulsed 400 Hz, 1 ms), 10-cm microwaves caused desynchronization of basal rhythms [31]. Low-voltage spiked waves from the motor-sensory region were noted. Exposure to CW at the same wavelength caused a similar phenomenon only after a temperature increase of 4.5°C was obtained. Exposures to 3-cm waves did not induce brain temperature increases or EEG changes. On the following day EEG tracings were normal. It should be pointed out that in all cases EEG registration was carried out after termination of irradiation.

In animals sacrificed immediately after exposure to 10-cm microwaves at 20 to 30 mW/cm^2, hyperemia of meninges, distension of superficial brain vessels, and small extravasations in deeper parts of the brain were seen. At lower power density no evident changes were found during macroscopic or microscopic examination. Also no changes were demonstrable in animals sacrificed on the day following exposure.

The power density used for repeated exposures 5 to 7 mW/cm^2, 3 h daily, 60 days caused a temperature increase of 0.5 to 1.0°C during a single session. In animals exposed to 10-cm PW, signs of desynchronization of the tracings obtained from the motor region became evident after 28 days exposures.

On the background of a desynchronic basal rhythm, among the periodically occurring slow waves (3 to 4/s) single sharp waves and spikes of 150- to 200-μV amplitude registered from the visual cortex region were observed. One month later the tracings became flattened and no dominant function could be discerned. Single

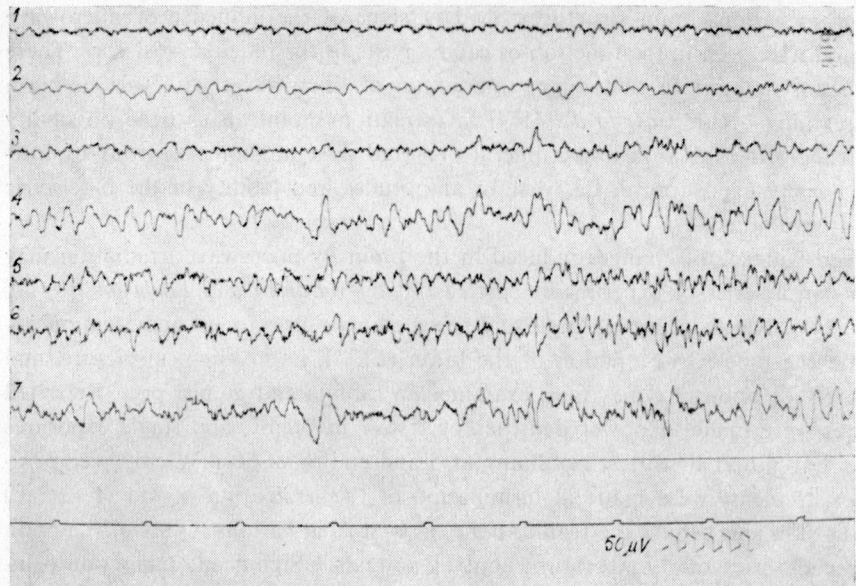

Fig. 49. Rabbit No. 162. Examined April 3, 1964. Examination was performed after 4-week irradiation with 10-cm wavelength and continuous modulation. The tracing shows no distinct abnormalities. Leads: 1, left motor sensory; 2, right motor sensory; 3, left sensory optic; 4, right sensory optic; 5, left motor, right motor; 6, left sensory, right sensory; 7, left optic, right optic.

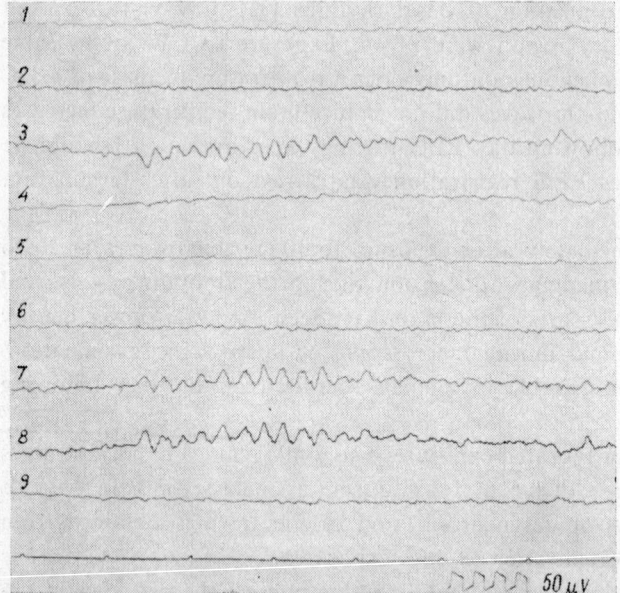

Fig. 50. The same rabbit as in Fig. 49. Examined May 22, 1964. The examination was performed after 7-week irradiation with 10-cm wavelength and continuous modulation. Distinct disorders of bioelectric activity. In the sensory-optic leads periodic series of waves of frequency 3 to 3.5 s of gradually rising and falling amplitude. Leads as in Fig. 49.

slow waves of increasing and decreasing amplitude occurred in the visual cortex region. Only traces of the normal reaction to light stimuli remained. Figure 49 shows a normal tracing of a rabbit and Fig. 50 a tracing after 7 weeks of irradiation.

Analogous exposure to CW 10-cm microwaves did not cause any differences in comparison with controls following 4 weeks of irradiation. After 60 days the amplitude of biopotentials decreased. Characteristic waves with a 3 to 4/s frequency and gradually increasing and decreasing amplitude were registered from the sensory-visual region. No EEG changes or morphologic signs of injury were found following exposures to 3-cm waves. Following repeated exposure to both CW and PW 10-cm microwaves, lesions were found on histologic examination of the brain. These lesions were more evident in animals exposed to PW irradiation. Nerve cells contained only sparse tigroid granules, a few were hyperchromatic, and signs of chronic Nissl disease were seen (Fig. 51). Spherical metachromatic (toluidine blue) structures were seen in the white matter of the brain and cerebellum. Histochemical analysis [19, 20] demonstrated the presence of proteins, mucopolysaccharides, and myelin in these structures. Their appearance is shown on Fig. 52. Phosphatides and cerebrosides were also present. The possibility that these structures could be amyloid bodies was excluded.

With the aim of analyzing the influence of microwave irradiation on particular brain structures, various drugs of known action were administered to irradiated animals [30, 142]. Luminal, phenactil, cardiasol, or strychnine was administered to animals following a single 10-cm PW microwave exposure to 20 mW/cm^2 or repeated 3 h/day exposures to 7 mW/cm^2, 70 to 80 h in total. A single exposure causes desynchronization of the bioelectric function of the brain in animals to which phenactil was administered prior to irradiation and decrease tolerance to cardiasol administration, but increases tolerance to strychnine. A single exposure after Luminal administration does not provoke any significant changes in the EEG.

Repeated exposures lead to a decrease of tolerance to cardiasol and strychnine administration as well as a reduction as the effects of phenactil administration. Administration of Luminal influences the cortical activity in chronically irradiated animals only to a slight degree. These observations may be interpreted as an indication that microwave irradiation influenced the cortex function (effects of Luminal plus irradiation) only to a very small extent. A comparison of the effects obtained by irradiation and phenactil or strychnine or cardiasol administration indicates that microwave irradiation influences the hypothalamic part of the reticular system. It may be recalled that morphologic evidence leads to a similar conclusion [20]. The general conclusion of the cited authors is that microwave irradiation may exert a stimulatory influence on the function of the reticular formation of the brain.

In another experiment [142] rabbits were irradiated for a total of 100 or 108 h (3 h daily, 10-cm PW, 7 mW/cm^2) and were administered chlorpromazine and/or D-tubocurarine, after which the EEG was recorded. Administration of D-tubocurarine led to the appearance of marked signs of excitation in the EEG, which may be

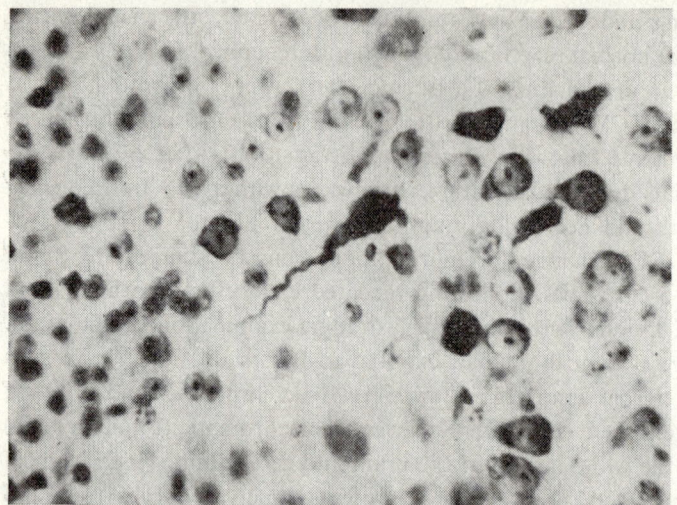

Fig. 51. Degeneration of cortical cells in the brain of a rabbit after 4-month exposure to microwaves. See the text. From [19].

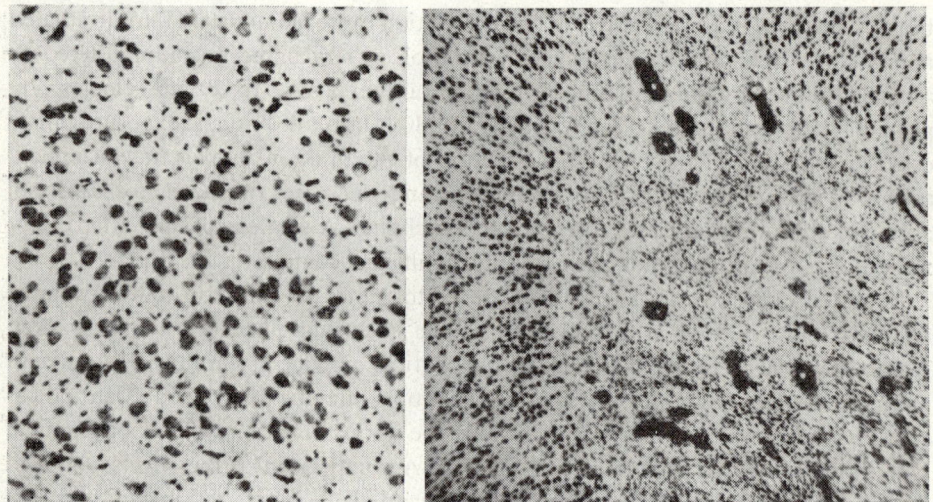

Fig. 52a.   *Fig. 52b.*

Fig. 52. Metachromatic bodies (*a*) and perivascular changes (*b*) in the brain of a rabbit after 4-month exposure to microwaves. See the text. From [19].

abolished by subsequent chlorpromazine injection. Tubocurarine administration before chlorpromazine injections does not counteract the synchronization of bioelectric brain function caused by chlorpromazine. These observations may be interpreted as an indication that synaptic structures at the brain-stem level are affected by microwave irradiation.

Servantie at al. [605, p. 36] independently found differences in the reaction of microwave-irradiated rats to pentatetrazol and curare-like drug administration. It should be stressed once more that pharmacodynamic analysis of microwave effects in the nervous and cardiovascular systems seems to be a most promising approach, which may permit clarification of many moot points concerning microwave interactions with these systems.

Another very promising approach are the quantitative studies of *Guy* and his associates, who determined the absorbed power density thresholds for microwave effects in the nervous and auditory systems of cats and humans (see [605, 606, 607]).

The reader interested in electroneurophysiologic research on microwave and radio-frequency effects should also acquaint himself with the paper by *Bawin*, *Gavalas-Medici* and *Adey* [608]. The data obtained by these authors indicate that low-level, very high frequency fields amplitude modulated at specific frequencies (4, 10, 13, 14 Hz) produce marked effects on conditioned specific brain rhythms (enhanced regularity of the patterns, sharpening of the spectral peaks around the central frequency of the response, and extremely prolonged resistance to extinction). The authors conclude that "a possibility exists that an externally applied electrical field could influence the activity or excitability of a population of cells and that the changes seen in gross recordings would reflect true neuronal phenomena."

It should be stressed that these observations were made using frequencies well outside the microwave range (147 MHz, amplitude modulated at 1 to 25 Hz). The results may be relevant, however, to such questions as the neurologic effects of various irradiation cycle rates.

Barański and *Edelwejn* [144] investigated the effect of a single exposure to 10-cm PW microwave at 20 mW/cm^2 on rabbits with electrodes implanted into the hypothalamic region of reticular formation and Amon's horn. The signs of excitation of these brain structures in EEG tracings recorded after the exposure may be the results of thermal or electrical stimulation by the electrodes during exposure [14, 209, 253].

The validity of all the cited finding of *Barański* and *Edelwejn* may be examined on similar grounds. In all experiments the animals underwent electrode implantation 1 week before exposure to microwaves and were irradiated with the electrodes present. These presumably distorted the configuration of the field. No thermal stimulation of deeper layers can be expected because the electrodes were fixed superficially in the external cortical layer of the parietal bone and reinforced by a layer of a plastic substance used in dentistry. Temperature measurements did not show any influence of the electrodes on temperature distribution. On the other hand, the surgery involved in electrode implantation influences the obtained EEG record, which becomes stable only 1 week after electrode implantation. No EEG recordings were made during irradiation, so no artifacts caused by interference of the microwave fields are to be expected. Moreover, in experiments on the effects of repeated low-dose exposures, initial (prior to irradiation) recordings did not differ from those made after 4 weeks of exposure, the differences appearing at a later period. No effects were obtained after exposure to 3-cm waves as opposed to the results of 10-cm irradiation. All these considerations seem to indicate that in the described experimental conditions microwave exposures influenced the bioelectric function of the brain in rabbits. It should be pointed out, however, that any general conclusions as to the quantitative relationship of exposure dose to bioeffect are not possible on the basis of these experiments. In addition, the conclusions based on electroencephalographic and pharmacodynamic studies remain in good accord with morphological and biochemical results [19, 20].

Table 17

Results of radioactivity measurements (imp/min/mg) of guinea pig brain in 10-cm-microwave-irradiated and control animals. Incubation with $DF^{32}P$; radioactivity indicates the relative amount of active acetylcholinesterase centers. According to [20]. See text.

Experiment	Mean ± standard deviation	Significance level
Control	1725 ± 103	
PW, 3 h, 25 mW/cm²	1125 ± 133	$p < 0.001$
CW, 3 h, 25 mW/cm²	1691 ± 204	$p < 0.05$
PW, 3 h, 3.5 mW/cm²	1351 ± 193	$p < 0.001$
CW, 3 h, 3.5 mW/cm²	1773 ± 360	$p > 0.7$
CW, repeated exposures	2082 ± 450	$p < 0.05$
PW, repeated exposures	2289 ± 240	$p < 0.001$

The histological findings in irradiated rabbits were mentioned previously. Examination of histologic sections of guinea pig brains following similar PW and CW 10-cm microwave exposure gave analogous results.

Diisopropyl fluorophosphate (DFP) tagged with ^{32}P was used for blocking the active centers of acetylcholinesterase (Ache). The relative quantity was compared on the basis of radioactivity determination of irradiated and nonirradiated animals.

Single exposure of guinea pigs at 3.5 and 25 mW/cm² to CW 10-cm microwaves did not influence Ache. Pulsed-wave irradiation caused a significant decrease as compared with controls. Repeated exposures both to PW and CW led to an increase in radioactivity in the brain, i.e., in the quantity of Ache. This phenomenon was more pronounced following PW exposures. Determinations of Ache in various parts of the brain, i.e., determinations of the radioactivity, were made. Single exposures (Table 18) led to a decrease and repeated exposures (Table 19) to an increase in Ache in guinea pigs. Opposite results were obtained in rabbits exposed to repeated

Table 18

Results of radioactivity measurements (imp/min/mg) in various parts of guinea pig brain following 3-h irradiation at 25 mW/cm^2 (10 cm). Incubation with DF^{32}P; radioactivity indicates the relative amount of active acetylcholinesterase centers [20]. See text.

Brain structure	Control Mean ± standard deviation	Experimental Mean ± standard deviation	Significance level
Cortex	825 ± 152	577 ± 160	$p < 0.01$
Diencephalon	1785 ± 304	1301 ± 290	$p < 0.01$
Mesencephalon	2035 ± 513	1606 ± 327	$p < 0.05$
Metencephalon	1502 ± 203	1463 ± 293	$p < 0.05$
Cerebellum	2854 ± 351	1812 ± 304	$p < 0.001$

Table 19

Same as Table 18, but following repeated chronic exposure; according to [20]. See text.

Brain structure	Control	PW (10 cm)
Cortex	1035	983
Diencephalon	1395	1627
Mesencephalon	1592	2147
Metencephalon	1338	2023
Cerebellum	1494	1661

Table 20

Radioactivity measurements indicating the relative amount of active acetylcholinesterase centers in the brain of rabbits exposed for 4 months to 10-cm microwaves (DF^{32}P, imp/min/mg). According to [20]. See text.

Experiment	Mean ± standard deviation	Significance level
Control	250 ± 20	
PW	260 ± 13.1	$p < 0.01$
CW	216 ± 20.7	$p < 0.02$

Table 21

Radioactivity measurements (imp/min/mg) indicating the relative amount of active acetylcholinesterase centers in various brain structures of rabbits irradiated for 4 months. According to [20]. See text.

Brain structure	Control	PW	CW
Cortex	171	167	165
Diencephalon	248	219	223
Mesencephalon	299	174	194
Metencephalon	314	271	282
Cerebellum	233	204	214

PW and CW microwaves in analogous conditions (Table 20). A significant decrease in Ache was observed, and more marked effects were obtained after PW irradiation (Table 21). It should be noted that in all experiments the midbrain part seems to be the most affected.

Succinic acid dehydrogenase (Table 22) and cytochrome oxidase activity (Table 23) were also determined. Differences in the results obtained after analogous exposures to CW or PW microwaves merit special attention. The effects obtained after

Table 22

Succinic acid dehydrogenase activity in the brain of guinea pigs following microwave irradiation; according to [20]. See text.

Experiment			Mean ± standard deviation	Significance level
		Control	29.5 ± 3.1	
Single exposure	PW	25 mW/cm^2	23.1 ± 0.96	$p < 0.001$
		3.5 mW/cm^2	24.2 ± 3.2	$p < 0.01$
	CW	25 mW/cm^2	34.2 ± 4.7	$p < 0.02$
		3.5 mW/cm^2	28.1 ± 2.2	$p < 0.05$
Repeated exposures		PW	26.1 ± 3.4	$p < 0.05$
		CW	31.5 ± 3.5	$p < 0.05$

Table 23

Cytochrome oxidase activity in the brain of guinea pigs following microwave irradiation; according to [20]. See text.

Experiment			Mean ± standard deviation	Significance level
		Control	262 ± 15.0	
Single exposure	PW	25 mW/cm^2	268 ± 13.0	$p < 0.05$
		3.5 mW/cm^2	272 ± 11.7	$p < 0.05$
	CW	25 mW/cm^2	308 ± 10.3	$p < 0.001$
		3.5 mW/cm^2	253 ± 10.1	$p < 0.05$
Repeated exposure		PW	236 ± 25.3	$p < 0.01$
		CW	257 ± 17.6	$p < 0.05$

single exposures to CW at higher power densities (20 mW/cm^2) may be explained as secondary temperature effects. The decrease of activity of the enzymes examined following repeated PW exposures and the differences between the effects of repeated PW and CW exposures are difficult to explain. It may be supposed that pulse peak power density influences the results. However, in the conditions of these experiments thermal (understood as temperature) effects should be exactly equal.

To obtain additional insight into the metabolism of the brain in irradiated animals, ^{35}S-sulfate and ^{32}P-phosphate were administered and the incorporation of these radioisotopes examined. The results obtained are presented in Tables 24, 27.

As seen in Table 24, no significant changes in ^{35}S incorporation occurred after irradiation. The remaining data show that ^{32}P incorporation decreased, especially following PW exposures. Tables 26 and 27 show that this may be explained by disturbances in the lipid and nucleoprotein metabolism.

The data obtained in this series of investigations have been discussed at some length. They represent a consistent attempt to characterize the effects of microwaves in the nervous system using various approaches and, in the present author's opinion, show the advisability of the simultaneous use of functional, morphologic, and biochemical methods for the investigation of microwave bioeffects.

Table 24

^{32}P and ^{35}S incorporation (imp/min/mg) into the brain of guinea pigs following repeated exposures to microwaves; according to [20]. See text.

Incorporation of:	Experiment		Mean ± standard deviation	Significance level
	^{32}P	Control	847.3 ± 96.2	
		CW	793.4 ± 71.5	$p < 0.05$
		PW	757.2 ± 53.7	$p < 0.02$
	^{35}S	Control	3784 ± 376	
		PW	3495 ± 667	$p < 0.05$

Table 25

Incorporation of ^{32}P (imp/min/mg) into the brain of rabbits following repeated exposures to microwaves; according to [20]. See text.

Experiment	Mean ± standard deviation	Significance level
Control	327.1 ± 32.4	
PW	239.0 ± 22.7	$p < 0.001$

Table 26

Specific radioactivity of guinea pig and rabbit brain fractions following repeated exposures to microwaves ^{32}P (imp/min/mg) [20].

Experiment	Fraction	Aqueous	Lipids	Nucleoproteins
Guinea pigs	Control	1548.1	1121.2	288.4
	CW	1862.0	989.9	258.2
	PW	1865.6	903.7	233.7
Rabbits	Control	2134.6	623.1	548.0
	PW	2862.6	544.5	476.9

Table 27

Index of ^{32}P turnover in the lipid and nucleoprotein fractions of guinea pig and rabbit brains following repeated exposures to microwave irradiation; according to [20]. See text.

Experiment		Lipid fraction	Nucleoprotein fraction
Guinea pigs	Control	72.4	18.6
	CW	53.2	13.9
	PW	48.4	12.5
Rabbits	Control	29.2	25.7
	PW	19.0	16.6

The preceding results are difficult to interpret; changes in acetylcholinesterase activity in irradiated animals have been described by many authors. *Nikogosjan* [404] found an initial increase and later a decrease of Ache activity in the blood of repeatedly irradiated rats (10 and 40 mW/cm², 10 cm). Irrradiation of rabbits for 2.5 or 5 months in similar conditions as in the experiments of *Barański* [20] gave similar results, a decrease in Ache activity. At a lower power density of 1 mW/cm² even an 8-month exposure period did not affect it. A marked decrease in Ache activity in the brain of rabbits irradiated with decimeter waves at 0.5 mW/cm² was reported by *Syngajevskaja et al.* [533]. In a further series of experiments the same author [530, 531] found a decrease in Ache activity in the blood accompanied by an increase of acetylcholine. The ATP and ADP content in the liver and skeletal and heart muscle was changed following microwave irradiation, which may be, according to *Syngajevskaja*, the result of disturbances in the cholinergic part of the vegetative nervous system. A decrease in Ache activity in the blood and brain of mice exposed to 10-cm microwaves was also found [7, 529].

several methods. Degenerative changes, disturbances in metabolism, etc., were observed in animals without implanted electrodes. It may be objected that similar metachromatic bodies (myelin degeneration) may be observed also in other conditions in the rabbits, e.g., under the influence of high-altitude hypoxia [18]; see [199]. Nevertheless, secondary effects of the action of various factors may be similar, in spite of differences in the primary mechanism. It seems that the pharmacodynamic analysis of microwave effects in the nervous system is a promising approach.

In view of these observations, a second general statement may be, formulated as follows:

Low-dose exposure at power densities of 1.5 and up to 30 mW/cm^2, according to certain authors, even at 0.2 and 0.5 mW/cm^2 [213], especially in the decimeter microwave range may induce changes in the bioelectric function of the brain in certain animal species. Prolonged periods of repeated exposures lead to progressive intensification of such changes. The results obtained by such types of experiments strongly point to the possibility that cumulative effects may be obtained. These effects seem to be secondary; the mechanism responsible for their appearance must be investigated further. Pharmacodynamic methods coupled with electroneurophysiological ones seem to offer a promising approach.

The numerous investigations cited demonstrate that low-power-density exposures 0.2 to 10 mW/cm^2 to 10-cm or lower-frequency microwaves may cause disturbances in conditioned-reflex function and induce abnormal behavioral effects. This was demonstrated in dogs, rats, and mice [193, 194, 316, 317, 319, 320, 512, 513]. The significance of these investigations was questioned by many authors (see discussions in [82] pp. 50, 134, 144, 179, 193). The present authors' impression is that the objections raised stem mainly from insufficient familiarity with the methods used in conditioned-reflex experiments. The presentation of this method lies outside the scope of this work. As the objections were raised mainly by English-speaking authors, a few fundamental works published in English are given in the references, including papers by *Pavlov* [419, 421], who established this method. It may be added that classical physiology textbooks, such as that of *Lovatt Evans*, contain chapters on this subject. *Bykov's* textbook of physiology, published in 1958 in London in English may be used as a useful source.

It should also be pointed out that the function of the nervous system may be considered on different levels, as a whole or on the cellular or molecular level. The first approach is of course fraught with difficulties, as our knowledge about the transmission, transformation, and integration of information in the nervous system is insufficient. It may be said that disturbances which may be detected using psychological, behavioral, or conditioned-reflex research methods are an indirect indication of microwave interference with the central nervous system. Such observations are purely empirical and may be classified as a phenomenological approach. Nevertheless, an empirical observation cannot be neglected or discarded simply because the operative mechanism is not understood. The behavioral effects of microwave exposure cannot be ignored [164, 165, 438]. The only legitimate approach is to note such facts and search for an explanation. It should be stressed, however, that extreme caution is needed before any definite conclusions are drawn.

The preceding considerations allow the following general statements to be made:

1. Exposure to microwaves, especially in the decimeter range, at power densities of about 1 mW/cm^2 (according to certain authors within the 0.2- to 10-mW/cm^2 power density range) may disturb normal unconditioned as well as conditioned reflexes in dogs [314-317, 512, 513, 525-527]. Such disturbances may be induced by whole-body exposure, irradiation of the head or its parts (remainder of the body screened or irradiation of the trunk with the head screened). Similar phenomena have been induced in rats [319-321, 382, 383] and mice [193, 194, 327]. In the majority of experiments the attempts to use microwave exposure as a conditioned stimulus did not succeed. The reports of *Justesen et al.* [270] and of *Malachov et al.* [327] are an exception. The latter obtained a negative defense reflex to 20-s irradiation with 2200-MHz waves at 20 mW/cm^2 after 30 to 50 exposures accompanied by an electroshock. Single exposures and repeated exposures at higher power densities (of the order of tens or hundreds of milliwatts per square centimeter) induce distinct effects.

2. Microwave exposures affect the behavior of various animals, mainly rats and mice [220, 256, 264, 269, 270, 280-282, 536]. The reports of *Kicovskaja* [264, 265], *Justesen et al.* [256, 270], and *Korbel* [280-282] merit special attention. Further data may be found in *Presman's* monograph [438].

3. The experiments of Adey and his group (608, 609, 611 see also 612) demonstrated that electromagnetic fields modulated at extreme low frequencies (8-16 Hz) interfere with calcium ion binding to the greater neuronal membrane. A quantum mechanical explanation of this phenomenon and its significance for the interpretation of EEG changes was offered by Grodsky (613).

It may be added that interesting observations on the influence of microwaves on birds have been made. As their physiology is different from that of mammals, the present authors do not feel competent to discuss these data. A chapter concerning this subject may be found in a recent monograph [141] and in the series of reports by *Tanner* and *Romero-Sierra* (see [573] for references). According to a personal communication (*Romero-Sierra*, [1972]) microwave exposures affect the development and productivity of eggs in hens; an increase in the incidence of fowl leukemia was also observed. The feathers of birds may serve as sensory detectors of microwaves [537].

The mechanism of microwave interaction with the nervous system are unclear. Solving this question seems to be of prime importance for the understanding of the primary and secondary effects of microwave exposure. A solution depends on advances in the knowledge of the physiology of the nervous system. Nevertheless, sufficient evidence exists to question the categorical statements of *Schwan*, e.g. [473,

474] that excitation of nerve cells is impossible in the light of "modern neurophysiology." *Schwan* restricts this opinion to human exposure; it should be pointed out, however, that many of the experiments cited were carried out in conditions comparable to those of human exposure during work with microwave equipment. *Schwan's* statements, based on theoretical considerations, seem to be contradicted by empirical observations. It should be kept in mind that at least a part of the phenomena observed may be secondary effects.

Various disturbances appear and become intensified in the course of prolonged periods of repeated, low-dose daily exposures. No explanation of the operative mechanism can be offered at this time.

One possible mode of induction of central nervous system effects is the secondary results of interaction with peripheral nerves and/or receptors. The existence of such mediated effects is demonstrated by changes in the EEG or conditioned reflexes following irradiation of peripheral body parts, the head being screened. Direct interaction with various brain structures, particularly with the reticular formation, at low power density levels that induce negligible temperature increases cannot be excluded. Although *Presman's* hypothesis (see [438] and Chapter 3) concerning the role of magnetic and electromagnetic fields in the transmission of biological information needs critical experimental verification, possible interference with such a physiological mechanism should be kept in mind. Finally, local and generalized cardiovascular reactions, endocrine reactions, and temperature-dependent or "nonthermal" metabolic changes may influence the central nervous system. In other words, the possibility exists that central nervous system effects are symptoms of interference with the functions of various organs, or of a stress imposed upon the regulatory and compensatory mechanisms of the organism.

Cardiovascular Effects of Microwave Exposure

The cardiovascular effects of microwave exposure at high power densities are the result of thermoregulatory compensatory reactions. These are put into action by mechanisms dependent, among others, on the nervous system, which in turn may be affected in its function by microwave-induced disturbances. The diversity of local vascular reactions points to such a possibility [196], as stated previously. Because of the interdependence of the cardiovascular and central nervous systems or peripheral nerve microwave-induced effects these aspects were discussed in the preceding section.

The circulatory effects induced by microwave hyperthermia of 40.5°C in rats obtained by 10-min irradiation with 2450-MHz waves at 80 mW/cm² were investigated by *Cooper, Pinakatt, Jellinek,* and *Richardson* [94-96, 428, 429, 456]. The animals were in pentobarbital anesthesia, which could influence the results. The blood pressure, heart-stroke volume, and beat rate increased, and the peripheral resistance decreased. A peak level was reached at the moment rectal temperature attained 40.5°C. Further temperature increases led to decompensation and death [95]. The cited authors examined the influence of various drugs on the circulatory reaction at its peak level, i.e., 40.5°C body temperature measured in the rectum. Pyridoxine or pyridoxal administration did not influence the blood pressure and heartbeat rate increases; heart-stroke volume remained at the initial level [93]. The mechanism of the effect produced by administration of these vitamins on the circulatory reaction is unclear. Digitoxin administration [429] did not influence the blood pressure and heartbeat rate increases, and prevented an increase in heart-stroke volume, although increasing it in control animals. *Ouabain* [428] abolished the circulatory reaction in irradiated animals, and increased the heart-stroke volume in control animals, while the blood pressure and heartbeat rate remained unchanged. These results are difficult to interpret. Reserpine administration [96], vagotomy, and pharmacologic ganglioplegia [94] diminished the reaction to microwave hyperthermia. Striking results were obtained following bilateral adrenalectomy, which completely abolishes any circulatory response to irradiation [94]. These results indicate the importance of the endocrine and vegetative mechanism in compensatory reactions to microwave hyperthermia (see [218]).

According to *Subbota* [511], the reaction to microwave-induced hyperthermia does not differ from responses observed in a hot environment (hot room) or following infrared irradiation. A single exposure to high power density levels causes an increase in blood pressure, heart-beat rate, and respiratory rate. Arrhythmia and ECG changes may appear [511]. Local heart irradiation at 100 to 200 mW/cm^2 may cause bradycardia and amplitude changes in the ECG (R, S and T waves), which are interpreted as a result of changes in calcium content [540, 541] in the blood.

Soviet authors have carried out numerous investigations on the influence of prolonged periods of repeated irradiations on the blood pressure in dogs, rabbits, and rats (e.g., [183, 406]). The results of these experiments were reviewed by *Subbota* [511, 514] and *Gordon* [185]. A short restatement of these findings, based to a great extent on both reviews, is given next.

Subbota [511] investigated the influence of 12.6-cm and decimeter (wavelength not specified) microwaves and infrared irradiations on blood pressure in rabbits. At 10 mW/cm^2 12.6-cm waves caused only slight variations in arterial blood pressure. Exposure to infrared, which induced a rectal temperature increase from 1 to 1.5°C, caused marked changes in blood pressure following the first irradiation sessions. After four to five sessions infrared irradiation did not induce any hemodynamic response. Exposure to 12.6-cm waves at 1 mW/cm^2 caused a change in the blood--pressure response of such infrared-adapted rabbits. On the following or even third day after microwave exposure, infrared irradiation caused a high (by about 20 mm Hg) blood-pressure rise. Another instance of this "desadaptive" influence of microwave exposure is the following experiment. Rabbits were exposed to 12.6-cm irradiations at 50 mW/cm^2 on 8 consecutive days; rectal temperature increased by 1 to 1.7°C.

Beginning from the fifth session no response in blood pressure was noted. On the eigth day irradiation with decimeter waves at 1 mW/cm^2 caused a blood-pressure rise of about 12 mm Hg; exposure to 12.6-cm waves on the following day caused a marked rise in blood pressure of about 12 mm Hg; the response to subsequent irradiation with decimeter waves 1 mW/cm^2 on the tenth day was lower, less than 10 mm Hg, and response to 12.6-cm waves at 50 mW/cm^2 on the eleventh day was about the same. Irradiations with decimeter waves at 1 mW/cm^2 followed directly by exposure to 12.6-cm waves at 50 mW/cm^2 on the twelfth to seventeenth days was accompanied by a very high blood pressure rise, about 35 mm Hg. Exposures to 12.6-cm waves at 1 mW/cm^2 at the beginning of the session, followed by an increase of power density to 50 mW/cm^2 on days 21 through 24 of the experiment caused only slight blood-pressure rises of a few millimeters. These experiments demonstrate differences in physiologic responses to superficial or deep body heating infrared and 12.6-cm irradiation, 12.6-cm and decimeter wave irradiation. *Subbota* bases his thesis on the desadaptive effects of microwave exposure among others on these and similar experiments (see also Chapter 3). It should be pointed out that microwave exposure does not provoke hemodynamic disturbances in rabbits adapted to a vertical position.

Gordon [183, 185] examined the blood pressure in rats exposed to repeated irradiations at various wavelengths during 6 to 8 months. The blood pressure rose following 10 mW/cm^2, 10-cm and decimeter-wave exposures during the first 22 and 10 weeks, respectively. Millimeter and 3-cm waves did not induce any response at this power density. Power densities of 40 to 100 mW/cm^2 were necessary to obtain a blood-pressure effect at these wavelengths. Following the initial period of increase (10-cm and decimeter waves), the blood pressure returned to normal and after this dropped below initial values. *Gordon* points out that different effects are obtained following exposures to 10-cm PW or CW; CW irradiation at 10 mW/cm^2 did not influence the blood pressure during the initial 8 weeks of exposure, following which a drop in blood pressure was noted. Ten-centimeter PW exposures caused an initial increase, followed by a return to normal and a subsequent decrease in blood pressure. *Abrikosov* [1] also noted differences in blood-pressure responses to PW or CW exposures. It should be pointed out that following such chronic experiments the blood pressure returned to normal 8 to 10 weeks after termination of the irradiation period. Soviet authors explain the above-described phenomena by the interaction of microwaves with the nervous system. In the present authors' opinion, *Selye's* concept of stress adaptation fatigue better fits these observations. A part of *Gordon's* results are shown in Table 28.

Nikonova [406] reports a monophasic blood-pressure response, which consisted of hypotensia, in rats exposed to 0.5-MHz waves 2 h daily for 10 months. Soviet authors stress the hypotensive effects of prolonged microwave and meter-wave exposure. This effect is ascribed to interference with the central nervous system function, cardiac function alterations being mediated through interaction with skin receptors.

Table 28

Arterial blood pressure changes in rats irradiated for 6 to 8 months with various wavelengths at various power densities. All results are significant on the $p = 0.02$ or 0.01 level; according to [185].

Band designation	Power density (mW/cm^2)	Mean increase during the initial phase (mm Hg)	Mean decrease during the final phase (mm Hg)
Millimeter	10		18.7
	40	22.9	19.4
3 cm	10		23.0
	100	10.0	25.2
10-cm PW	1	12.7	14.7
	10	8.4	9.4
10-cm CW	10		8.1
Decimeter	1		7.2
	10	7.9	17.5

Presman and *Levitina* [311, 312, 445, 446] examined the influence on rabbits of 10-cm, PW or CW, and 12.5-cm microwaves on the heart-beat rate. At low and medium power density of 7 to 12 mW/cm^2 CW or 3 to 5 mW/cm^2 PW (pulse width 1 ms, repetition rate 700 Hz), exposure to 10-cm-wave irradiation of ventral parts of the body caused a slowing down of the heart-beat rate; irradiation of the head or the vertebral column caused acceleration. The decrease in heart-beat rate was observed immediately during the exposure; acceleration was observed at its termination. The effects of PW exposure were more marked than those of CW irradiation. In a further series of experiments the animals were exposed to CW irradiation for 20 min two times per second during 0.1 s at 700 to 1200 mW/cm^2 or to a pulse series (1 ms, 700 Hz) at mean power density of 350 to 380 mW/cm^2. The exposure of any arbitrarily selected region of the body caused a deceleration of the heartbeat rate. This effect could be abolished by mesocaine skin anesthesia. The authors conclude that the acceleration of the heart-beat rate following head and dorsal spinal cord exposure is caused by microwave interaction at the central nervous system level. The effects of exposure of the ventral side of the body are the result of skin-receptor excitation. This is claimed to be the dominant effect at exposure to high power densities. In the present authors' opinion, thermal effects in peripheral nerves or central nervous system neurites similar to those described by *McAfee* (p. 114) could also play a role. No effects on heart-beat rate were observed following exposures at 1 mW/cm^2.

Kaplan et al. [261] repeated a part of *Presman* and *Levitina*'s experiments [445] using analogous equipment and an experimental arrangement based on *Presman's* description [439]. *Kaplan et al.* [261] irradiated rabbits with 12.5-cm (2409 MHz) continuous waves during 20 min at 10 mW/cm^2, as did *Presman* and *Levitina*. It should be stressed that the former placed the animals in the far-field zone, the latter in the near-field zone. In spite of the contention of *Kaplan et al.* that this should not influence the obtained results, the present authors' doubts about the comparability of exposure conditions are not dispelled, see [564] and Chapters 2 and 3. Dorsal parts of the head of 16 rabbits divided arbitrally into four groups were irradiated. The results obtained in group 1 could be interpreted as heart-beat rate acceleration during exposure and at its termination. In group 2 a slowing down of heart action during the initial 10 min of exposure and a return to normal during the next 10 min were demonstrated. In group 3 deceleration during an 8-min period followed by a slight acceleration were noted. In group 4 an acceleration of the heartbeat rate was seen during the fourth, sixth and twenty-sixth minutes of the observation period. All groups were examined under identical exposure conditions, and a statistical analysis of the variations observed in all 16 animals indicates that all changes in heartbeat rate are not chance variations. According to *Kaplan et al.* [261], the results obtained by *Presman* and *Levitina* are based on too small a number of animals examined in each experimental variant; the observed changes may be discounted as random variations in the heart-beat rate. *Kaplan et al.* examined only CW irradiation effects, citing both parts of *Presman* and *Levitina*'s report, i.e., concerning CW and PW exposure effects. It should be noted that the latter were more pronounced. It seems that further critical experiments are necessary, the more so as others report heart-beat rate effects due to microwave exposure, which could be correlated with EEG and morphologic changes in the brain of rate irradiated at 5 to 15 mW/cm^2 [247]. *Nikonova* [406]; see also p. 119, correlated blood-pressure changes in rates with electric- or magnetic-field intensities, claiming that the latter play a more important role. *Tjagin* [552] reported ECG changes in dogs during low-dose, 5- to 10-mW/cm^2, partial-body exposures.

Observations on microwave exposure effects on heart action in frogs further complicate the picture. Whole-body exposure may cause a decrease in the heartbeat rate, irradiation of the head, an increase. No effects are obtained by direct irradiation of the denervated heart *in situ* [313].

Frey [164], see also [170], observed the dependence of effects obtained by 1425-MHz PW exposure on the time relationship between the phase of heart action and the moment of pulse action. Frog heart-beat rate may be accelerated if the pulse occurs about 200 ms after the P deflection, i.e., simultaneously with QRS. In about a half of the experiments, arrhythmia, followed sometimes by heart arrest, was observed. If the pulse occurs simultaneous with the P deflection, no arrhythmia may be induced. It may be provoked, however, if the pulse occurs concomitantly with the R deflection. *Paff et al.* [414, 415] described ECG changes in chicken embryonic heart action on irradiation with 24,000 MHz at 74, 167, and 478 mW/cm^2

for periods of a few seconds to 3 min at 38°C. Changes in S and T deflections and a shortening of the QT wave and heartbeat rate were observed.

The available data on cardiovascular responses to microwave irradiation may be summed up as follows:

1. Exposure at power density levels that cause temperature increases induces a hemodynamic response related to the thermoregulatory compensatory reaction. The role played by the central and/or vegetative nervous system as well as the humoral (endocrine) stress reaction of the hypophyseal — adrenal system must be more closely investigated, as differences between the hemodynamic reaction induced by microwave heating and heating by other means occur (see [94, 95, 196]).

2. Cardiovascular responses to low-dose and PW exposure merit further investigations; of particular interest are experiments designed to examine the dependence of the effects obtained on the time relationship between the phase of heart bioelectric action and pulse occurrence [164, 170].

3. Cholinergic, adrenergic, modern cardiovascular drugs, and various inhibitors (blockers) could be applied usefully for an exploration of the mechanisms involved in cardiovascular responses to microwave exposure. The existing reports are based on experiments with anesthetized animals, which may invalidate conclusions drawn from the obtained results. No recent papers on this subject were found.

Endocrine and Metabolic Effects of Microwave Exposure

In spite of the obvious importance of this subject, only sparse and not always sufficiently documented data were found. The discussion of endocrine effects will be restricted to the thyroid gland and adrenals; the influence of microwaves on gonads is presented in the next section.

The effects described in the preceding sections indicate that responses to high power density exposures on part of the hypothalamic-hypophyseal-adrenal system are to be expected. It suffices to recall only hemodynamic responses [363]. No satisfactory investigations on this aspect of microwave effects were found. The examination of the possible effects of repeated low and median exposures on the hypothalamic-hypophyseal system seems to be of prime importance. A close examination of the morphologic description and photograms in the monograph by *Tolgskaja* and *Gordon* [554] show that the hypothalamic region of the brain is affected; discrete changes in cells and synaptic connections are detectable. The same authors [555] recently presented evidence that the above-described phases of deviations in blood pressure in rats subjected to prolonged periods of low-dose repeated exposures may be correlated with changes in the amount of neurosecretion detectable on histological sections. This seems to support *Mikołajczyk's* contention that microwave exposure affects the hypothalamus, the effects on the function of the hypophysis being secondary. *Mikołajczyk* [368, 370] administered hypophyses from normal and microwave (10 cm, CW) irradiated adult rats to young hypophysectomized rats. In both groups the adrenals increased in weight normally. The inference is that microwave irradiation does not affect the amount of adrenotropic hormones in the hypophysis. The same author reports, however, that the amount of gonadotropic hormones in male and female rat hypophyses increases slightly after microwave irradiation and decreases 18 h after the session (10-cm CW, at 0.01, 1, 3, 10, 20 and 150 mW/cm^2, 1 h/day, single or repeated exposures) at power density levels from 10 mW/cm^2. Microwave irradiation is thought by *Mikolajczyk* to affect hypothalamic factors governing FSH and LH release from the hypophysis [370].

More data are available on the influence of microwave exposures on adrenal function. *Leńko et al.* [305] reported a decrease of 17-OCHS in the urine of rabbits during the initial period of 20 exposures on consecutive days (4 h/day, 50 to 60 mW/cm^2, 10-cm CW). During the later period beginning from the tenth day, the 17-OCHS level returned to normal. This may be related to the previously described phenomena of adaptation to repeated microwave exposures. No changes were found in the 17-CS urine excretion. *Mikolajczyk* [370] did not find any changes in corticosteroid content in the adrenals and blood serum of rate exposed to 10-cm CW microwaves at 10 mW/cm^2 for 15, 30, or 60 min.

Leites and *Skurichina* [304] exposed rats to 10-cm microwaves at 100 mW/cm^2 for 10 min and examined the histological picture of the adrenals during the following 14 days. Initially, a decrease occurred in Sudan III positive lipids, birefringent substances, and vitamin C content. Following this, an increase as compared with controls was observed, and the picture returned to normal after 2 weeks. Similar observations were made by *Gorodeckaja* [192] after a 5-min exposure to 3-cm waves at 400 mW/cm^2.

Syngajevskaja et al. [533] exposed large groups of dogs and rabbits (162 animals) to decimeter waves. Changes in glucose levels in the blood, deviations in hyperglycemia curves following sugar administration, and variations in liver glycogen content were noted. Lactic acid levels were also affected. The authors suggest that disturbances in carbohydrate metabolism are mediated through effects on the hypophyseal—adrenal system, as well as on the liver and pancreas. In view of the complexity of factors influencing carbohydrate metabolism, the mechanism of microwave effects remains open. *Barański et al.* [29] reported changes in pyruvic acid and lactic acid blood levels, accompanied by a decrease in glycogen content in skeletal muscles in rabbits following 10-cm irradiation at 5 mW/cm^2. The simultaneous increase of multiphasic potentials and prolongation of the duration of single and multiphasic potentials in EMG tracings led these authors to the conclusion that the changes found may be related to metabolic effects in skeletal muscles.

In view of these observations the disturbances in carbohydrate metabolism induced by microwave irradiation should be studied more closely.

Nikogosjan [405] described lesions in the adrenal cortex following microwave exposure. Repeated prolonged exposures of rats to decimeter waves at 40 mW/cm^2

1 h daily did not influence the sodium and potassium levels in the blood [291]. Disturbances in mineral balance should be expected if any serious adrenal cortex lesions should be present (mineralocorticosteroid secretion). An increase of calcium levels in the blood and in urine was reported [291], which may be related to another observation [358], that of chloride level increase in the blood.

Several reports indicate that microwave irradiations many influence the thyroid gland; disturbances in its function were observed following the exposure of dogs to 2800- and 1240-MHz microwave [237, 239]. Radioiodine uptake by the thyroid gland was augmented for 4 to 25 days following exposure 6 h daily for 6 consecutive days to 1240-MHz PW 360-Hz repetition rate, 2-ms pulse width at 50 mW/cm^2 [237]. The same paper contains a rather disquieting observation that radioiodine uptake was augmented in comparison to control values in dogs 3 to 4 years after a single exposure to 1280 MHz at 100 mW/cm^2. The authors explain these changes as secondary effects of influence on the function of the hypothalamic—hypophyseal system. *Barański et al.* [32] also report an increase in radioiodine uptake, as well as histologic and electronomicroscopic signs of thyroid hyperfunction in rabbits exposed repeatedly to 10-cm waves at 5 mW/cm^2.

These observations indicate that various metabolic disturbances may be expected as the results of microwave interference with endocrine function. It should be kept in mind that temperature effects influence the metabolism and may lead to displacements in water and mineral equilibrium [359]. These will be reflected in the peripheral blood, which suggests unexplained, complex metabolic disturbances. It should also be stressed that many isolated reports on various metabolic effects exist (see [425] and [438]). It is difficult, however, to find a logical pattern allowing postulations concerning any general mechanisms responsible for the observed disturbances. It is to be regretted that no systematic, controlled investigations on a well-known chain of metabolic reactions in microwave-irradiated animals were found in the surveyed literature.

Oxygen consumption and carbon dioxide production in 10-cm-microwave-irradiated rate could be correlated with incident power density values and the resulting body temperature increase [123]. Carbon dioxide levels decrease in venous blood in dogs exposed to 1240-MHz microwaves at 50 mW/cm^2 [358], which may be explained by changes in the respiratory rate accompanying the thermoregulatory response. Oxygen content in the arterial blood of dogs following 2800-MHz irradiation was also lower than in controls; respiratory distress was reported in dogs and rabbits irradiated with 2800 MHz and rats irradiated with 2450 or 10,000 MHz [356, 452]. All these phenomena seem to depend on body-temperature increases.

Microwave irradiation lowers the resistance to high-altitude hypoxia or hypoxia caused by inhalation of gas mixtures of low oxygen content 11 and 8.5 percent. In such conditions microwave-irradiated animals responded with a marked decrease in red cell counts in blood, instead of the usual hypoxic polycythemia [423].

Several authors reported changes in blood serum proteins following microwave exposure. Following repeated exposures to 1240-MHz waves, *Michaelson et al.* [358] found a decrease in total serum protein levels in dogs. *Nikogosjan* [405] reports changes in the relation of albumins to globulins from 1.6 to 1.1 in rats following 10 to 20 daily 1-h exposures to centimeter and decimeter waves at 10 mW/cm^2; further more 20 to 40 exposures cause a decrease in alpha- and betaglobulins. *Gelfan* and *Sadtchikova* [175] report an increase in globulin, which is responsible for changes in the albumin-globulin index. *Singatulina* [498] observed a decrease in serum albumins in rabbits following prolonged periods of repeated exposures to meter waves. *Syngajevskaja* and *Sinenko* (cited according to [529]) found, after a 30-min exposure of dogs to meter waves at 3 mW/cm^2, an increase in serum blood levels of cystine, lysine, arginine, glutamine, glycine, glutaminic acid, tryptophan and phenylalanine, and a decrease in the tyrosine and leucine levels. Similar changes were seen in rabbits [529]. *Gruszecki* [202, 203] reports an increase of alpha-2- and betaglobulins accompanied by gammaglobulin decrease in rabbits exposed to pulsed 10- and 3-cm microwaves for 80 h at very low power densities of 5 or 10 mW/cm^2. Glutamine-oxalo-acetic transaminase levels were about two times higher than in normal animals. The alkali reserve fell to 18.5—20 mEq/liter as compared to the normal value of 27 mEq/liter. Both the description of the experimental conditions and a personal discussion with the author indicate that the animals were placed in the near-field zone. The values of power density stated in this report may be questioned and the actual exposure was probably higher. Variation in the albumin-globulin index in the blood serum of rabbits following single or repeated exposures to 3- or 10-cm microwaves at power densities of 5 to 25 mW/cm^2 were also reported [523, 524]. Soviet authors explain these observations as a result of interference with liver or adrenal function [529], this last supposition being based on the fact that corticosteroids influence amino acid metabolism. However, disturbances in protein metabolism may be related to changes in protein turnover in other organs. ^{35}S-methionine and ^{32}P-phosphate turnover alterations were reported to occur in the central nervous system of repeatedly irradiated rabbits [20] and in the testes [248] and ovaries [231] of rats exposed to lower frequencies. *Miro et al.* [389] found an increase in ^{35}S-methionine incorporation in the liver, spleen and thymus of mice exposed for 145 h to pulsed 3105-MHz microwaves at 20-mW/cm^2 mean power density and 400-mW/cm^2 peak power density. All these results indicate that further controlled experiments are needed to obtain a picture of the microwave effects on protein metabolism in various organs and in the organism as a whole system. It should also be realized that temperature effects influence such results to a great extent; adequate data on temperature control were not found in the cited reports.

Sacchitelli and *Sacchitelli* [463, 464] reported alterations in the glutathione level and activity of amylase and lipase in the blood of guinea pigs and rats exposed to 2500-MHz microwaves. Changes in DNA or RNA content in the skin and internal organs in microwave-irradiated rats were also described. *Kerova* [262] found an increase in nucleic acid content in rats exposed for 6 min to 3-cm microwaves at

100 mW/cm² or 500 mW/cm², accompanied by a decrease of nuclease activity. *Syngajevskaja et al.* [532, 533] found an increase in DNA and RNA content in the spleen and a decrease in the skin of rats exposed to millimeter waves at 200 mW/cm² for 30 min which caused a rectal temperature increase of 0.5 °C.

Syngajevskaja carried out a series of investigations on ATP, ADP, and phosphorus levels in various organs of microwave-irradiated animals. The findings obtained are cited here according to a review published in 1970 [529]; see also [530]. The starting point for these investigations was observation on alterations of oxidoreductive processes in microwave-irradiated animals. Exposure of rats to 3-cm microwaves at 30 to 49 mW/cm² for 30 min caused an increase in rectal temperature of 0.7 to 1.2 °C, accompanied by a decrease of succinic dehydrogenase and cytochrome oxidase activities in the heart muscle, liver, and kidneys. Further increase of the power density to 150 — 170 mW/cm² intensified this effect. Irradiation of rabbits with 10-cm waves at the same power density caused a marked decrease in the activity of these enzymes. At higher power densities the activity of ATPase in the kidneys, liver, and muscles of rabbits increases (*Moskaljuk*, cited according to [529]). In view of this, *Syngajevskaja* began to investigate the ATP, ADP, and inorganic phosphorus content in the liver, heart muscle, skeletal muscles, and brain of rats irradiated with 10-cm microwaves for 1 h at 10 mW/cm² and 5 min at 100 mW/cm². With exposure at the lower power density, ATP content decreased. ADP and inorganic phosphorus levels increased. The activity of dehydrogenase and cytochrome oxidase was higher than in controls; the activity of ATPase was unchanged. At the higher power density the rectal temperature rose to 40-40.5 °C; ATP and the otal ATP plus ADP, as well as inorganic phosphorus levels, in the liver, skeletalt muscles, and brain increased. In the heart muscle the ATP and ADP content was lowered; inorganic phosphorus increased.

It is extremely difficult to sum up this evidence. All the aspects of microwave interaction presented in this section need further investigations, both on the course and dose dependence of the phenomena described and the mechanism involved. Alterations of the hypophysis, adrenals, and thyroid gland function seem to be well documented. The interdependence of these alterations should be examined, as well as the influence of microwave irradiation on the hypothalamic-hypophyseal system. It should be noted that morphologic observations have demonstrated changes in the pineal body [76]. Numerous reports indicate that microwave exposure induces disturbances in protein metabolism, carbohydrate metabolism, and oxidoreductive processes. Such observations should be correlated with disturbances in endocrine function; their dependence on temperature effects must be clarified and quantified. It may also be noted that microwave irradiation offers an interesting experimental model for investigations of local and generalized temperature elevation. A comparison of the metabolic effects of hyperpyrexia induced by various means with microwave effects may contribute to an understanding of the mechanisms involved.

Influence of Microwave Irradiation on Testes, Female Genital System, Pregnancy, and Foetal Development

Imig et al., in 1948 [243] were the first to draw attention to testicular effects of microwave irradiation. These authors examined the effect of scrotum exposure to 2450-MHz microwave or infrared radiation. Irradiation led to testicular enlargement, edema, degenerative changes in seminiferous vesicles, and fibrosis. Microwave irradiation induced such changes at lower temperature than infrared radiation. This was interpreted by many as evidence of additional "extrathermal effects." Nevertheless, testicular lesions were induced in these experiments only at high power densities. Similar results were obtained by *Cieciura* and *Minecki* [74, 77–79] following single or repeated exposures of rats to 2860-MHz CW microwaves at 60 to 90 mW/cm^2 for 1 to 2 min. The animals were sacrificed 24 h, 1, 4 and 6 weeks after exposure. Hemorrhages, degeneration of testicular vesicles, and abnormal naturation of spermatozoa were observed. Microwave exposure led also to a decrease in alkaline phosphatase, acid phosphatase, ATPase, succinic acid dehydrogenase, and 5-nucleotidase activities, as demonstrated by histochemical methods. The temperature of the scrotum and testicles was elevated in these experiments to 41.5 °C. Hypothermia was used for investigation of the role of temperature increase [79]. Ice-water was used for cooling the animals to 24 °C; 10-cm CW microwave irradiation for 10 min at 90 mW/cm^2 served to raise the temperature to normal levels. Infrared-irradiated hypothermic animals were used as controls. No changes were demonstrated in the control group. In microwave-irradiated animals, 24 h after and on the seventh day after the exposure several lesions were found on histologic examination. The blood vessels were distended, and hyperemia and edema were present (serous exudate in perivascular spaces and a few seminiferous vesicles). One week after exposure, the lumen of seminiferous vesicles was smaller than in controls. Alterations in the normal sequence of maturation phases of spermatozoa were found. No abnormalities were demonstrated in the spermatogonia; the number of spermatocytes was diminished, and spermatides, particularly the early stages, were absent. Irregular groups of spermatozoa forming thick aggregates were found in the lumen of seminiferous vesicles. An additional group of animals was subjected to testicular heating to 65 °C by immersion of the scrotum in a hot paraffin bath. The findings in this group were analogous to those in the hypothermic microwave irradiated group. The authors controlled the temperature using thermistor elements and an electric thermometer, Ellab, with a sensitivity of 0.1°C. The authors (see also [378]) conclude that microwave-induced testicular effects depend on "additional factors" besides temperature increase. In the present authors' opinion a general doubt may be raised concerning the comparability of superficial heating effects (infrared, paraffin bath) and deep, volume heating with microwaves. Evidently, such objections would apply to any experiment of this type.

Gunn et al. [205, 206] made similar observations following 15-min exposures to 24,000-MHz waves at 250 mW/cm². Superficial scrotum burns and temperature elevations to 45°C were observed. Twenty-nine days after the exposure, degeneration of the seminiferous epithelium and testicular fibrosis occurred. Shortening the irradiation period to 10 min caused a diminution of the intensity of observed lesions. Five-minute exposure caused only edema; the temperature did not exceed 41°C. The authors examined also the testicular endocrine secretory function by making use of ^{65}Zn incorporation into the prostate from the androgen level. After 15- and 10-min exposures, ^{65}Zn incorporation was markedly lower than normal. Five-minute exposures caused a decrease in incorporation by 45 percent. No changes in ^{65}Zn incorporation into the prostate occurred after infrared heating of the testes to 41°C. According to *Gunn et al.* [205], this may indicate that additional effects besides temperature effects are obtained by microwave energy absorption. *Searle et al.* [486] were also able to demonstrate differences in effects obtained by microwave or infrared irradiation at identical temperatures. Conversely *Imig et al.* [243] dit not find any differences between the effects of microwave or infrared irradiation and the surgical displacement of testes into the abdominal cavity. *Michaelson* [353] in reviewing the testicular effects of 2450-, 3000-, 10,000- and 24,000-MHz exposures of rats, rabbits, and dogs at power densities of 10 to 15 mW/cm², states that edema, atrophy, fibrosis, and necrotic areas were observed in seminiferous vesicles. According to this author, all these changes should be ascribed to temperature effects. Investigations on the influence of testicular heating by other means seem to support this opinion [147, 486]. A decrease of ^{35}S-methionine incorporation into the testes of microwave-irradiated hamsters [248] may also be a temperature-dependent phenomenon. *Ely et al.* [150] established the threshold value for testicular damage in dogs exposed to 2880 MHz at more than 10 mW/cm² for unlimited exposure time.

Additional interesting observations on possible testicular damage during radar equipment operation were reported by *Dolatowski et al.* [132, 306]. Rabbits were placed for 2 to 3 months in the far-field zone of intended radiation emitted by a 3-cm radar; the mean power density was about 0.3 mW/cm². No signs of testicular damage were found in this group. Involution of Sertoli cells was found in rabbits placed in the vicinity of the magnetron in the transmitter part, where the intended microwave radiation was below 10 mW/cm². The authors point to the nonintended soft X-ray generation as the possible cause of testicular damage. In the present authors' opinion, undetected nonintended microwave radiation could also play a role. Involution of seminiferous vesicle epithelium was reported also in mice exposed to prolonged microwave irradiation by *Prausnitz* and *Susskind* [437].

The preceding findings are difficult to correlate with existing specific reports on the absence of any adverse effects on mating behavior or offspring of microwave-irradiated animals. These reports concern dogs exposed to 24,000 MHz [117]; guinea pigs, to 3000 MHz [161]; and rats and mice, 3000 MHz [388] single or a few repeated irradiations. Adverse effects were, however, observed by *Gorodeckaja* [193]. White outbred mice were exposed to 3-cm PW microwaves for 5 min at 400 mW/cm^2 (repetition rate 577 Hz, peak nominal power 60 kW). Animals placed for 15 min in a hot chamber at 52°C served as controls. A part of the animals was sacrificed for histologic examination of the testicles or ovaries. The remaining animals were mated and their offspring observed.

Testicular lesions analogous to those described previously were found on histologic examination. The number of newborn animals in litters fathered during the first 5 days following irradiation was significantly decreased; an increase in stillbirths was noted. The development of offspring of irradiated males was normal. No mention is made of inborn defects. Ten days after the irradiation full normalization occurred.

Lesions of Graaf's follicles were found in ovaries of irradiated females. The total number of offspring and the number of animals born in each litter decreased more distinctly than in the case of the offspring of irradiated males. Stillbirths were frequent. Again no mention of inborn defects is made. No systematic lesions were found on histologic examination of the uterus; signs of damage to ovaries were present on examination 20 days after the irradiation. The development of the offspring of irradiated females was retarded, as demonstrated by slow gain in body weight. Analogous findings were obtained in control animals exposed to a hot environment (hot chamber). Histological signs of testicular damage were less evident; no effects on the offspring were noted.

Gorodeckaja [193] also observed disturbances in the sexual cycle of irradiated females. A more detailed investigation on this aspect was carried out by *Łętowski* [323, 324]. Female rats were irradiated with PW 2980-MHz microwaves at power densities of 0.1 and 1.3 mW/cm^2 for 4 h daily for 62 to 80 days. The sexual cycle lengthened, and signs of the rut period were poorly expressed. The cytologic picture of vaginal smears was altered. Low doses of radiation caused an increase in cytoplasmic RNA content; high doses caused a decrease. The DNA content was unchanged. No significant changes were found in the endometrium of irradiated animals. Venous congestion and proliferation of stromal cells were seen in the ovaries, accompanied by a decrease in RNA content in sexual epithelium cells. *Povzhitkov et al.* [435] exposed female mice to 3-cm PW microwaves at 0.344 mW/cm^2 for 30 min 20 or 50 times before mating or on 12 consecutive days for 10 min at 50 mW/cm^2 after mating. No deviations from normal were observed in the course of pregnancy, number of offspring in litters, or development of newborn animals in the group of females irradiated before mating. Irradiation during the first half of pregnancy caused its prolongation, retarded the development of offspring, and caused only a slight increase in stillbirths.

It may be mentioned that on March 1, 1972, in US Medicine an interview (Microwaves Inhibit Tumor Induction, by *R. W. Rhein*, p. 22) with *D. R. Justensen* appeared, in which it was claimed that fetal irradiation on the eleventh through the fifteenth days of gestation influences in mice the development of transplantable lymphoreticular cell sarcoma, inoculated on the sixteenth postpartum day. These observations are incomplete. In the paper quoted it is stressed expressly that the observations have "hypothesis value", and, last but not least, the report does not come from the investigator himself.

The influence of microwave irradiation on fetal development has been investigated mainly on embryo chickens. The first observations made by *Van Everdingen* [577, 578] indicated that exposures to 1875-MHz microwaves during the initial 5 days of development induce metabolic disturbances and death. Irradiation at later periods, on the eleventh day, did not induce any such effects. Effects on embryonic heart action and ECG were reported (see the section on cardiovascular-effects and [414, 415]). At certain power density levels injury and death of the embryo may be obtained; in the majority of experiments thermal temperature effects seem to play a decisive role [578, 583], ECG effects excepted [414, 415]. *Osborne* did not find any effects from 200-MHz exposure on embryo chick development; the temperature increased no more than 1°C following irradiation [410]. *Kondra et al.* [276] also report negative observations; hatching and development were normal following irradiation. Similar irradiation of hen eggs at various power densities with 3- or 10-cm microwaves carried out in our Institute did not bring conclusive results; any observed changes were attributable to temperature variations. The study was discontinued as not being contributive to the problems of microwave interaction with living systems (unpublished results.)

Minecki [381] cooled hen eggs to 16°C and warmed them to 37°C by subsequent irradiation with 10-cm CW microwaves at 90 mW/cm^2 during 2.5 min. Single exposures before incubation, after 24 h of incubation, and on the tenth day did not produce any effects. Daily exposures during 10 days of incubation led to a decrease of hatched chicken to 29.6 percent; in a control group subjected to analogous temperature variations (in an incubator), 92.3 percent of the eggs were hatched, as compared to 73 percent in the control group. According to the author, this may indicate that additional mechanisms, besides secondary temperature effects, are operative. The difference between superficial and deep heating should be once more pointed out.

The preceding experiments on the influence of microwave radiation on chick embryo development were conducted using dipole or horn antennas. A more controlled arrangement was used by *Van Ummersen* [583]. Eggs were introduced into the terminal section of a waveguide, and mismatches were tuned out. Low-power irradiation for prolonged periods (5 h), causing a yolk temperature increase to 42.5°C, induced abnormal development — hind limbs, tail, and heart defects; abnormal development of brain and eyes. An incubator-temperature-controlled series did not show any of the abnormalities found in the irradiated embryos. The author concluded that additional effects, besides the temperature dependent ones, occur. Taking these observations as a starting point, *Carpenter* and *Livstone* [69] irradiated pupae of the mealworm beetle, *Tenebrio molitor*, in a similar wave guide arrangement at 10.155 MHz. Power densities of 80 mW/cm^2 for 20 or 30 min or 20 mW/cm^2 for 120 min were used. Temperature was controlled at various points of the body and no "hot" spots were found to occur. Pupal deaths or abnormal development were induced by microwave irradiation. The abnormalities consisted in the absence, reduction, or shredding of wings or wing covers. The temperature during 20-mW/cm^2 exposures increased by 3°C; temperature-controlled series without irradiation controlled temperature chambers were examined. In the temperature-controlled series, normal development was observed in 17 of 20 pupae. In the microwave-irradiated series (20 mW/cm^2 for 120 min) 5 out of 25 pupae developed normally. Similar proportions of abnormalities were observed in insect pupae exposed to microwave irradiation for other periods or at other power densities. According to the authors, these experiments prove definitely the existence of additional effects that cannot be correlated with temperature increases. It would be extremely interesting to pursue investigations along such lines. The process of metamorphosis in several insects is relatively well known and depends on hormonal regulation. In the present authors' opinion this particular experimental model offers several promising possibilities.

Rugh and coworkers (in [605, p. 98]) demonstrated experimentally teratogenic microwave effects in mice. It should be stressed that the effects obtained seem to depend on the developmental stage and do not show any specific traits depending on a specific action of microwaves. The mechanism and possible significance of these effects for human health protection measures should be investigated further.

It seems that the possibility of teratogenic effects is sufficiently well documented if pregnant animals are exposed in given conditions. It should be stressed that studies on the effects on meiosis, particularly during spermatogenesis, are badly needed.

In summing up this section it must be stressed that more controlled experiments are needed. At present it may be stated only that exposure to power densities at levels causing temperature increases results in testicular lesions, particularly damage to spermatogenesis, in experimental animals. These lesions seem to be readily reversible, with the exception of extended thermal necrosis. Similar damage to the female genital system, particularly the ovaries, may be expected. These phenomena may lead to changes in fertility and an increase in stillbirths. The present authors' feeling is that no serious effects are to be expected at power density levels below 10 mW/cm^2 under usual exposure conditions.

As concerns the effects on fetal development, temperature effects seem to play the most important role. Very early and early stages of development are more sensitive, and microwave exposure at these stages may cause serious injury. Irradiation of insects at various developmental stages may serve as a convenient model for investigations on the mechanisms of microwave interaction with living systems. The applicability of these results to the interpretation of phenomena observed in mammals remains to be demonstrated. The obvious importance of the questions raised in this section makes the need for further controlled critical experiments imperative.

Chromosomal Effects, Possible Genetic Effects, and Influence of Microwave Radiation on Mitosis. Cellular Effects

Chromosomal aberrations and mitotic abnormalities may be induced, at least under certain conditions and in certain cell types, by exposure to microwave or radiofrequency fields. This is a well-established fact, as several reports from at least five independent laboratories exist. A series of investigations was carried out by *Heller* and his coworkers at the New England Institute for Medical Research, Ridgefield, Connecticut. These authors were the first to observe such effects in 1958, and have continued their research till now, see [221] for a review. The impact of these investigations comes from the observation that radio-frequency fields 5 to 7 MHz cause the migration of motile *Protozoa* with the electric field lines or across the field 27 to 30 MHz. Intracellular organelles showed similar frequency-dependent orientation, see [544] and Chapter 3, pearl chain formation. Chromosomal and mitotic aberrations could be induced in garlic root tips exposed to 5-40 MHz PW at various pulse duration 15 to 30 ms, repetition rates 500 to 1000 Hz and intensities varying between 250 to 6000 V/cm. Bridging, fragmentation, micronuclei, and chromosome lagging were observed [221, 222, 365]. 21-MHz exposures caused the most pronounced effects. Similar effects were obtained with human lymphocytes and Chinese hamster lung cells cultivated *in vitro*. Analysis of metaphasal plates demonstrated translocations, fragmentations, ring forms, and dicentrics. It should be stressed that such chromosomal aberrations are typical for ionizing radiation injury. Further studies 21-MHz PW with similar electric field intensities demonstrated that mutations, both sex-linked and autosomal, and crossing over in males (!) could be induced in irradiated *Drosophila* flies [221, 365]. It is interesting to note that the dormancy period could be broken in gladiolus bulbs, and more vigorous plants were obtained [222]. It may also be mentioned that, depending on exposure conditions the degeneration or acceleration of the growth of young tomato plants can be obtained at low frequencies (*Kucia*, personal communication, 1972).

These studies concerned the effects of irradiation well outside the microwave range. *Yao* and *Jiles* [593] obtained chromatid breaks, isochromatid breaks, dicentrics and rings, and chromatid exchanges in choroid and bone marrow rat kangaroo cells irradiated *in vitro* with 2450-MHz CW microwaves at 200 mW/cm^2, i.e., 50 cm from the antenna. Longer periods of irradiation up to 30 min and nearer the antenna (10 cm from it) caused cell death; exposures further from the antenna and for shorter periods of 10 min depressed cell proliferation and induced the chromosome aberrations mentioned. Special attention is merited by specific chromosome changes, that give the impression of partial despiralization. Such a picture is never seen in ionizing radiation induced damage. Similar "electromagnetic deteriorated chromosomes" were detected by *Stodolnik-Barańska* [506] in microphotograms of chromosomes of human lymphocytes irradiated *in vitro* with 10-cm waves at power densities of 7 to 14 mW/cm² up to three repeated exposures. The same author [507] obtained in these conditions the activation of human lymphocytes *in vitro* and the initiation of what is called lymphoblastoid transformation.

In should be pointed out that quantitation of the exposure of the irradiated cells in the last three experiments is extremely difficult. The field intensity within the flasks used for the culture may be only a matter of conjecture. *Stodolnik-Barańska* [506] controlled the temperature of the medium; no rises above 38°C were noted. No temperature measurements are reported by *Yao* and *Jiles* [593]. A preliminary attempt at quantitation was made by *Czerski* [100], who used a wave-guide arrangement similar to that of *Carpenter* and *Livstone* [69]. Human lymphocyte cultures were irradiated with 3-cm microwaves at various power densities and for different periods; the temperature was controlled. If was shown that at power densities below 3 mW/cm² and above 20 mW/cm² only heating and cell death could be obtained.

As the irradiated containers were not cooled, prolonged exposure to the low-level doses used in this experiment caused heating after a given time, after which simple thermal death of the exposed cells ensued. If the irradiation was interrupted earlier, just before overheating (over 38.5°C) occurred, no effects were observed; particularly, no blastoid transformation could be induced at levels below 3 mW/cm².

Exposures at power densities between 5 and 15 mW/cm², if continued till the moment when the temperature of the medium attained 38°C, could induce lymphoblastoid transformation; no such phenomenon could be obtained by exposure below 5 mW/cm². This problem merits further study; the human lymphocyte cultured *in vitro* may be considered a promising model for investigation of microwave interaction at the cellular level.

Chromosomal aberrations and changes in the duration of particular phases of mitosis (mitotic abnormalities) were reported in human lymphocyte cultures and cultures of monkey kidney cells following exposures at 3 mW/cm² to 7 mW/cm² to 10-cm PW and CW microwaves [28]. Mitotic abnormalities similar to those obtained *in vitro* may be observed in cells of the blood-forming system *in vivo* [320] (see also the section on the blood-forming system).

It should also be mentioned that changes in mitoses of *Vicia faba* root tips similar to those described in garlic root tips were obtained by irradiation with 10-cm microwaves [75].

Another interesting phenomenon is that microwave irradiation at low power densities of 10 mW/cm² 10-cm wavelength, changes the circadian rhythm of the epithelial corneal cell mitoses [369]. After irradiation with 24,500-MHz CW cataractogenic doses, an initial depression of DNA synthesis in the lens, followed by a precipitous rise on the fourth to fifth day after irradiation, was reported in rabbits [585]. Microwave exposure affects also the circadian rhythm of bone marrow cell mitoses (see the section on the blood-forming system).

Giant mast cell formation *in vivo* following irradiation with microwaves was reported [582]. Suppression of metachromatic staining and ^{35}S-methionine incorporation into these cells was obtained by *in vitro* irradiation (10 cm CW, 3 mW/cm^2, up to 10 min by *Sawicki* and *Ostrowski* [471]). It may also be mentioned that PW 9600-MHz irradiation (0.25-ms pulse width, 400-Hz repetition rate) affected the respiration and metabolism of guinea pig skin tissue culture [62, 63, 296, 297].

All these reports contain isolated observations and it is difficult to correlate the findings and obtain a reasonable pattern. At present it may be said only that sufficient evidence exists to postulate the possibility of microwave and radio-frequency irradiation-induced effects at the cellular and subcellular level, which cannot reasonably be correlated with temperature effects. It seems that approaches stemming from *Debye's* theory are insufficient to explain these phenomena. Molecular effects that influence the spatial structure, energy levels, and, in consequence, the biological and metabolic activity of macromolecules may be responsible. This possibility must be carefully examined and cannot be contradicted or categorically confirmed before more data are accumulated. Nevertheless, the probability of such mechanism being involved, according to the present authors' feeling, is rather high, particularly in view of chromosomal alterations suggesting despiralization.

It remains only to add that no satisfactory evidence of microwave-induced genetic effects or fetal damage in mammals was presented. It has even been suggested that microwave thawing of deep-frozen semen of cattle may be a better method than water-bath thawing [39]. *Bereznickaja et al.* [42] investigated the effects of low-dose, 10-cm microwave exposure on several generations of white outbred mice. A decrease in fertility and slower body weight gains in young animals were described. No definite abnormalities or inborn genetic defects were found by these authors.

Sigler et al. [497] suggested that mongolism (Down's syndrome; trisomy 21) occurs more frequently among the offspring of fathers professionally exposed to microwaves. The statistical significance of these data is doubtful. Once more turning to lower-frequency radiation effects, it may be mentioned that *Harmsen* (according to [336]) found an increase in the percentage of females among the offspring of rats exposed to 3- to 4-m radiowaves. Similar observations were made by *Manczarski* [329], who observed a prevalence of females among the offspring of radio-station or television-transmitter personnel meter and medium radio waves. The statistical significance of these data is extremely doubtful. The observation is mentioned because of its persistence, both in the literature and the apprehensions voiced by the personnel concerned. It may be added that in microwave-irradiated rats a prevalence of male offspring was noted (*Bereznickaja*, personal communication, 1972).

Nevertheless, extreme caution must be exercised. A few cases of multiple inborn defects in the offspring of women irradiated immediately before or during the early stages of pregnancy with shortwave diathermy have been described. One such case was reported by *Coccorza et al.* [83], a second by *Erdman* [155], and three are cited by *Minecki* [375]. These reports are difficult to interpret; because of the serious consequences involved, they should be carefully noted by the physiotherapist. Moreover, certain observations on magnetic field effects (see [33]), indicate caution. The genetic effects described in *Drosophila* are also disquieting.

Microwave Effects on Internal Organs: Chest, Abdominal Cavity, and Digestive Tract

Only a few critically controlled experiments on this subject were reported in the surveyed literature. Numerous papers exist in which various macroscopic and microscopic lesions seen on the autopsy of animals exposed to particular microwave bands at different power densities are described [77, 188, 191, 193, 234, 249, 326, 379, 403, 487, 488, 554, 556, 557, 600, 601]. All these reports contain more or less detailed descriptions of degenerative lesions, which can be explained by focal hyperthermic effects or vascular injury. Such lesions are seen most frequently in the liver [326, 600], in certain instances in the kidneys [309], or in the spleen. The effects on splenic tissue will be discussed in the next section on the blood-forming system. Focal lesions in the heart muscle were also described. Similar changes were found in the nervous system. Several authors point out that lesions and hemorrhagic foci may be found in the adrenal glands.

Studies of this type are of restricted value for the understanding of microwave bioeffects. Only selected data will be presented; a few representative macroscopic and histologic findings will be cited, which can be condensed to the following comments:

1. The distribution of focal lesions corresponds to the inhomogeneity of microwave field distribution and energy absorption within the body. The position of the animal in relation to the plane wave front (idealized far-field conditions) influences the results; different changes may be expected if the animal is irradiated frontally, laterally, ventrally, or dorsally.

2. The resulting lesions seem to depend not only upon vascularization of the tissue (organ) but also upon the reaction of blood vessels to irradiation.

3. Certain data indicate that the local vascular reaction to irradiation may vary with the even at identical power densities [196]. The reaction of peripheral blood vessels to microwave exposure and the mechanism involved in this reaction should be more closely investigated: the role of generalized hormonal responses, the role of nervous system responses, the reaction of intravascular receptors, and microwave effects in contactile elements.

4. Microwave effects on liver function should be examined; the majority of authors found morphologic signs of injury to this organ. Disturbances in carbohydrate metabolism and blood enzyme activity indicate disturbances in liver function. The

reports on the differences in the effects of microwave irradiation on Malpighian corpuscles and renal tubules need verification.

Only a few papers on the influence of microwave irradiation on the digestive tract have been published. *Grebieshnikova* [198] found retardation of water passage to further parts of the tract in dogs with ventricular or duodenal fistulas following exposures to decimeter or meter waves at power densities of 0.6 to 1.0 mW/cm^2. In control animals the water remained in the stomach for 10 min; in irradiated ones, for 50 to 60 min. In guinea pigs the passage of contrasting substances through the digestive tract is retarded, i.e., peristaltic movements are slower [198, 548]. *Fajtelberg-Blank* [158, 159] obtained different results. Irradiation of the ventricular region with meter waves in dogs with the "small stomach", according to Pavlov (surgical displacement of a part of the stomach outside the body, a variation of the ventricular fistula used for ventricular function investigation), accelerated water passage and absorption of chlorides and amino acids.

Subbota [509] observed retardation of food passage from the stomach in dogs following whole-body exposures to 10-cm waves at 1 mW/cm^2 for 30 min or decimeter waves at 0.5 mW/cm^2 1 h. This effect becomes less distinct on repeated exposures, and finally may be absent. After partial denervation all nerves around the cardia were cut through, irradiation causes the opposite effects, i.e., stomach movements are accelerated. Ventricular contractions become less frequent and have a longer duration in a fasting animal following irradiation.

The secondary function of the stomach is also affected. Following decimeter- or meter-wave irradiations at 1 mW/cm^2, the total amount of gastric juice, its total acidity, and free hydrochloric acid were decreased. The first phase (neuroreflectory) of the secretory function is the most affected. Repeated irradiations have a decreasing effect.

Subbota [509, 514] pays much attention to the effects of high-power-density exposure. Ventral exposure of the epigastric region of rabbits to 12.6-cm waves at 100 to 120 mW/cm^2 for 10 min causes the appearance of ulcers on the ventral wall of the stomach in about half the animals. The temperature of water in a balloon introduced into the stomach did not rise over 40 to 42°C during temperature-control exposures. An increase of power density to 150 mW/cm^2 caused the induction of ulcers in nearly all animals. The ulcers appear on the third day following exposure. Necrosis of the mucous membrane, frequently of the submucosa and only rarely in deeper layers, i.e., in the gastric muscles was seen. Irradiation of caudal parts of the abdominal wall induced similar ulcers in the small and/or large intestine. In all cases the ulcers were well delineated, limited focal lesions. After an interval of 10 to 20 days the appearance of ventricular ulcers corresponded closely to that of venticular ulcers in man [430].

Subbota [509] also examined urinary excretion from the left and right kidney in dogs with ureters led out to the surface of the abdomen and irradiated with microwaves. The animals were administered 500 ml of water with milk and the amount the excreted urine was measured. Lateral irradiation during 30 min with 10-cm waves at 1 to 5 mW/cm² caused a decrease in urine excretion on the irradiated side and an increase on the opposite side. Similar results were obtained following lateral exposure to decimeter waves at 0.5 mW/cm². An opposite phenomenon occurred on the next day; after irradiation the excretion was increased on the irradiated side and decreased on the opposite side. This effect became less evident on subsequent irradiations. These results indicate that vascular reactions to microwave exposure may occur in the kidney following low-dose irradiation.

It is difficult to attempt an evaluation of these observations of the influence of whole-body exposure to microwave radiation on the function of the digestive tract and kidneys. The results point to the possibility of disturbances in secretion (excretion) and absorption, which may be related to vascular responses and changes in blood flow. Disturbances in peristaltic movements merit attention. All such disturbances may cause secondary effects. Actual data are insufficient and constitute only isolated phenomenological observations. Further investigations based on determinations of microwave energy absorption distribution are necessary.

Microwave Effects on Blood and the Blood-Forming System

Conflicting reports on the effects of microwave exposure on the peripheral blood picture exist. The differences among the results obtained may be ascribed to species specifity and/or various exposure conditions. Certain reports are based on small numbers of individuals. *Deichmann et al.* [117] describe the effects on two dogs of 20-month exposure to 24,000-MHz radiation PW at 20 mW/cm² for a total of 2631 and 3970 h. No changes in blood volume, hematocrit, RBC, WBC, and differential blood count were noted; the only symptom attributable to irradiation was a slight body-weight loss. Similarly, *Spalding et al.* [504] did not find any differences between control, irradiated RFM-strain mice 24 animals exposed to 800 MHz at 42 mW/cm² for 2 h daily, 5 days per week and for 35 weeks in the peripheral blood picture, body weight, voluntary activity, and mean life span. One wonders about the significance of this study, as the irradiated objects are about one half-wavelength in size. *Prausnitz* and *Susskind* [436]; see also [437]; irradiated 100 mice for a period of 59 weeks, 4.5 min daily at 100 mW/cm² with 3.2-cm microwaves. An increase in WBC occurred, accompanied by lymphocytosis in the differential count. Longevity of irradiated animals was slightly affected. The most disquieting finding was the occurrence of leukemia in 35 percent of the irradiated animals as compared to 10 percent in the control group. *Kicovskaja* [263] irradiated white rats with 10-cm microwaves at 10, 40, and 100 mW/cm² 1 h daily for 216 days. No changes were found in the first group; at higher power densities a slight decrease in RBC, as well lymphocytosis or lymphopenia, was seen. An increase in RBC was noted in rats exposed to millimeter waves [188].

A series of investigations on the hematologic responses in rabbits and dogs was carried out by *Michaelson et al.* [238, 357, 360, 362, 363, 549]. The animals were exposed to PW 1240-, 2800-, or 200-MHz microwaves at power densities of 50, 100, or 165 mW/cm^2. These authors used dogs for comprehensive studies on microwave influence on the hematopoietic system. Briefly, these results indicate that displacements of white blood cells may occur immediately after exposure, as expressed by lymphopenia and eosinophilia. It may be noted here that *Szmigielski* [535] demonstrated that microwave irradiation on single or repeated exposures does not influence the production of granulocytes in rabbits but changes their distribution within the body, i.e., the relationships among the bone marrow granulocyte reserve pool, the marginal, and the circulating granulocyte pools (10 cm, 5 to 7 mW/cm^2).

Michaelson et al. demonstrated differences in peripheral WBC changes that depended on wavelength and/or power density. These phenomena may be explained by variations in granulocyte distribution among the various body pools due to thermal effects, i.e., vascular reactions and adrenal response to the thermal stress. Changes in lymphocyte and eosinophil counts make such an explanation probable.

An interesting finding is the shortening of red-cell survival time, as demonstrated by ^{51}Cr tagging of erythrocytes. A discrete hemolysis is claimed to be responsible. Discrete signs of changes in blood volume and hematocrit values were also reported. Iron metabolism was also affected; nevertheless, ferrokinetic studies did produce equivocal results. ^{59}Fe incorporation into erythrocytes rose, and plasma iron clearance was increased or decreased, as well as plasma iron turnover. In two dogs examined following prolonged repeated exposures a total of 138 and 60 h, plasma iron clearance was prolonged and erythrocyte iron incorporation increased. Hematocrit values, RBC, and hemoglobulin level fell.

One of the most interesting findings of these authors is that microwave exposures may affect the sensitivity to ionizing radiation. Exposure to 100 mW/cm^2 irradiation with 2800-MHz PW microwave decreased the mortality from subsequent X-ray exposure. Microwave exposure following X-ray irradiation or simultaneous exposure caused adverse effects. The same phenomenon or lack of effect was observed following microwave exposures at 165 mW/cm^2. The mechanism of the influence of microwave radiation on the course of ionizing radiation-sickness syndrome and its hematologic symptoms are difficult to explain. The authors conclude [237] that microwave exposure results in alterations of the compensatory and homeokinetic mechanism of the body. It should be noted that microwave exposure following X-ray irradiation may reveal the residual radiation damage [238]. This effect was studied and confirmed by Soviet authors, not only as concerns X-ray effects, but also with reference to the influence of other injurious factors [423]. These effects are termed by *Subbota*, see Chapter 3, as the desadaptive effect of microwave irradiation.

It may also be interesting to note that exposures of *Escherichia coli* cultures to 3105-MHz FW microwaves, pulse duration 1 μs, repetition rate 50 Hz, peak power of the transmitter 250 KW, mean power 12 to 13 W, i.e., peak power density 38 W/cm^2, mean 2 mW/cm^2 had a definite radio-protective effect, which was even more expressed in the second, non-microwave-irradiated generation. This single observation of *Miro et al.* [391] indicates that genetic (molecular?) effects may be involved in the radio-protection afforded by microwave radiation. Both because of the practical implications and importance for the understanding of microwave bio-effects, studies along these lines should be pursued further.

Hematologic effects of microwave low-dose exposures repeated for prolonged periods were studied extensively by Polish authors. These studies have been partly published in English [22, 102], but for the most part they are available only in Polish and will be presented in more detail.

Gruszecki [202, 203] found only slight variations in the peripheral blood of rabbits exposed to microwaves below 1 mW/cm^2 power density. *Barański* [20, 22] examined systematically the peripheral blood picture of guinea pigs and rabbits exposed to 10-cm PW or CW microwaves 3 h daily at 3.5 or 5 to 7 mW/cm^2 for various periods 2 or 3 weeks and 3 to 4 months. No significant changes were observed in RBC; a slight tendency to hypochromic anemia developed in rabbits after 4-month exposure. A marked increase in WBC was observed in rabbits and guinea pigs after prolonged periods more than 2 months and in all animals after 3 months. Differential counts demonstrated that this increase was caused solely by an augmentation of absolute lymphocyte counts; more than twofold increases were noted. After termination of the experiment, a further slight increase was noted, up to 2 weeks, followed by a slow return to normal values (Fig. 53).

Examination of lymph nodes and spleen demonstrated the presence of blastic forms, mitoses, and increased thymidyne incorporation (autoradiographic studies), which shows that many cells of the lymphoid system actively synthetized DNA.

In this connection the investigation of *Miro et al.* [389] should be recalled. Swiss albino mice were exposed to 3105-MHz microwaves at a mean power density of 20 mW/cm^2 peak pulse power density 400 W/cm^2 for 145 h. In the spleen, signs of

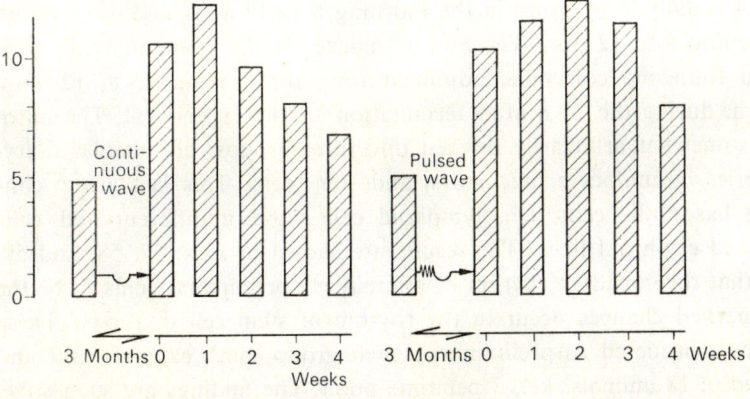

Fig. 53. Absolute lymphocyte counts before (first column) and after CW or PW irradiation for 3 months and during 4 weeks after exposures. See the text. From [22].

stimulation of lymphopoiesis were found on histologic examination, as well as increased ^{35}S-methionine incorporation in germinative centers and surrounding lymphoid cells. Similarly, an increase of ^{35}S-methionine incorporation into the thymus and liver cells was noted. The authors interpret these results as a sign of stimulation of cells belonging to the reticulo-endothelial system.

Only slight variations were found in bone marrow counts by *Barański*. Perhaps the most interesting findings in this study were mitotic abnormalities, similar to those described in garlic and bean root tips (see p. 132). Changes in the nuclear structure of erythroblasts and cells of the lymphoid system which consisted of fragmentation, presence of micronuclei, and unequal staining of chromatin, were observed. Bridges between daughter nuclei and chromosome lagging were also observed. It is interesting to note that such changes were seen only in the erythroblasts and lymphoid cells. Mitoses in the granulocytic precursor cell line were normal and no aberrations were noted [28]. The cytologic picture of affected bone marrow and lymphoid cells is shown in Figs. 54, 55, and 56.

It seems that, at least in these exposure conditions, no significant findings, the absolute lymphocyte count excepted, are obtained using routine hematologic methods. Qualitative cytologic analysis should be pursued further, however, as the described phenomena may be dependent on molecular changes that determine the chromosome structure [221].

Janes et al. [248] did not find chromosomal aberrations in the bone marrow of irradiated Chinese hamsters. Stickiness of the chromosomes, which could lead to aberrations in the second celluar generation, was found to occur in a significant percentage of examined mitoses. It should be stressed that the animals were exposed to 12-min irradiation in a microwave oven (2450 MHz) at rather high field intensities.

Another observation [27] indicates that the cricadian rhythm of bone marrow cell divisions may be altered by microwave exposure. Guinea pigs were exposed to 10-cm CW microwaves in an anechoic chamber in the far-field zone to about 1 mW/cm^2 for 4 h daily, one group in the morning 8 to 12 a.m. and the second group in the evening 8 to 12 p.m. The mitotic indexes in the bone marrow were then determined following colcemide administration at 12 p.m., 4, 8, 12 a.m. and 4 and 8 p.m. during the 24 h after termination of the experiment. The mitotic forms were somewhat arbitrarily divided into three groups: erythroblasts, erythrocytic cell series, granulocyte precursors, and stem cells. This last group consisted of young basophilic cells, i.e., lymphoid cells, and undifferentiated cells of the myeloid and erythroid lines. The results are shown in Figs. 57, 58, and 59. It may be seen that the circadian rhythm of granulocyte precursors seems to be the least affected; marked changes occur in the rhythm of stem cell divisions. These results should be considered as preliminary; each group (both experimental and control) consisted of 18 animals, i.e., 3 per time point. The findings are suggestive, but no statistical evaluation is possible. Further experiments demonstrated that a single exposure of inbred DBA mice may cause similar changes. Low-power-density

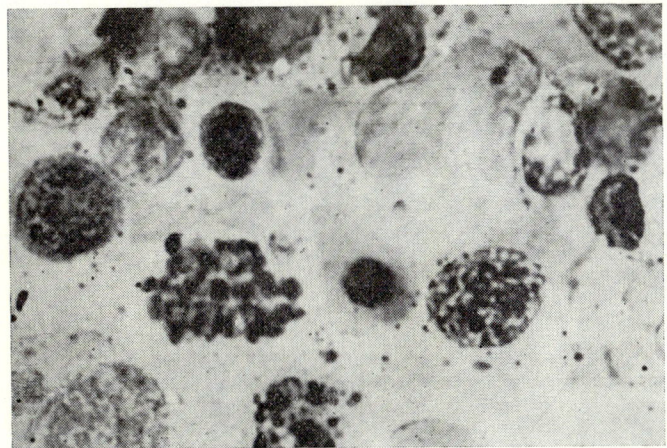

Fig. 54. Metaphase plate with abnormal chromosome structure (hyperpycnotic regions) in a lymph node imprint preparation. See the text. From [22].

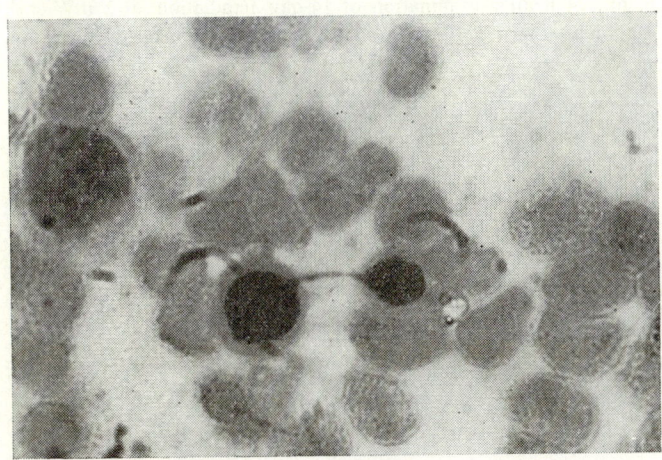

Fig. 55. Chromatin bridge between two daughter erythroblasts (bone marrow). See the text. From [22].

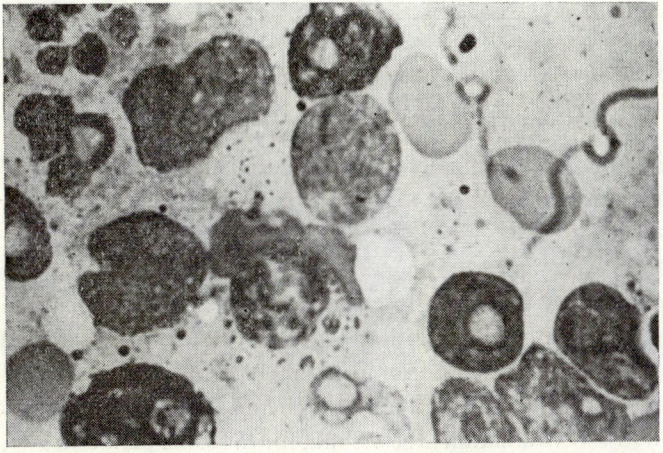

Fig. 56. Nuclear chromatin structure abnormalities in a spleen imprint preparation. See the text. From [22].

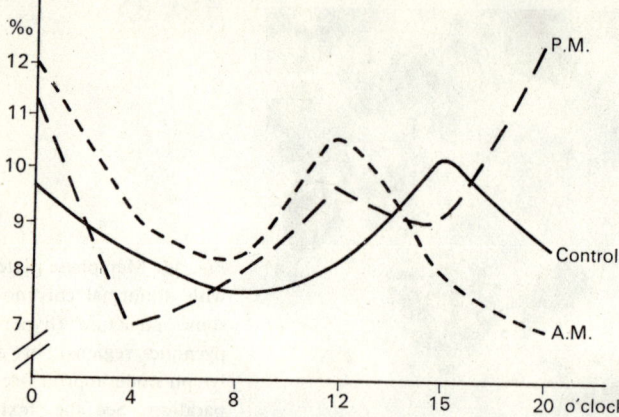

Fig. 57. Circadian rhythm of the mitotic index of erythroblasts in guinea-pig bone marrow. Control, nonirradiated animals; p. m., 24 h after termination of 14-day irradiation at 1 mW/cm², 10-cm PW, 8 to 12 p. m. – a. m., 24 h after termination of 14-day irradiation but from 8 to 12 a. m. From [27].

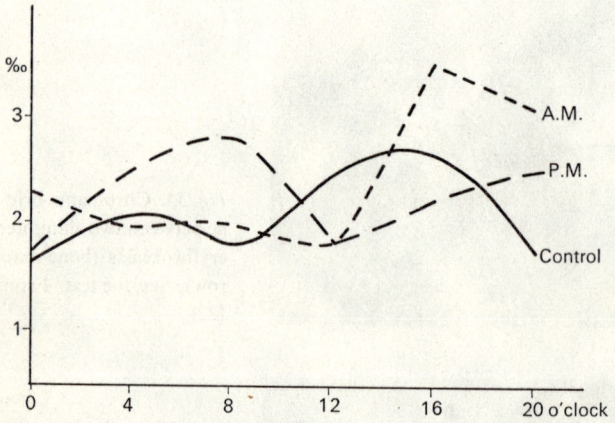

Fig. 58. Circadian rhythm of the mitotic index of granulocyte precursors in the bone marrow of irradiated and nonirradiated guinea pigs. See Fig. 57. From [27].

exposure (less than 1 mW/cm²) to 10-cm microwaves induced disturbances in the circadian rhythm of bone marrow mitoses for 48 h after irradiation; on the third day the rhythm returned to normal [614].

All the preceding findings were more pronounced after exposure to PW 400-Hz repetition rate, 1-s pulse duration than CW 10-cm microwaves. This difference is best demonstrated by the findings of *Siekierzyński* [101, 495, 496], who examined iron metabolism using ^{59}Fe (ferrokinetic studies) in 10-cm PW and CW irradiated rabbits. The animals were exposed for 2 h daily at 3 mW/cm². The first group was exposed for a total of 74 h to CW irradiation; the second, for the same period to PW,

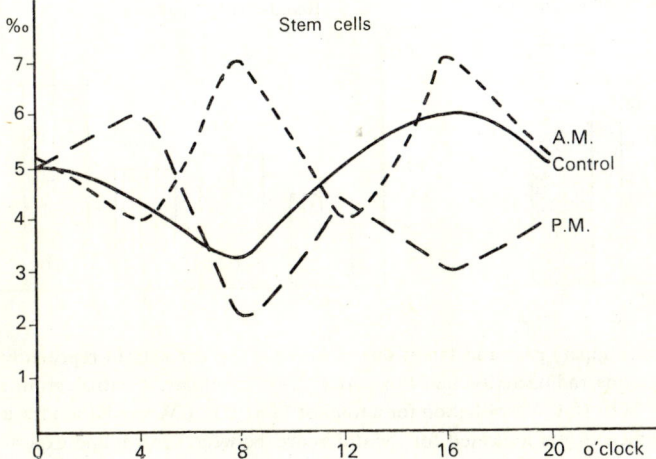

Fig. 59. Circadian rhythm of the mitotic index of "stem cells" in the bone marrow of irradiated and nonirradiated guinea pigs. See Fig. 57. From [27].

and the third, to 158 h of CW. The RBC, hemoglobin level, and hematocrit did not change significantly. Ferrokinetic studies demonstrated, however, that iron metabolism was affected and erythrocyte production, as determined by ^{59}Fe incorporation, decreased. This phenomenon did not express self in RBC count probably because of too short period of observation. The results obtained are shown in Figs. 60–63. It should be noted that CW exposure did not cause marked effects after a total of 74 h; significant findings were obtained, however, after an identical period of exposure to PW irradiation. It is also interesting to note that exposure to CW during a total 158 h induced effects similar to those obtained by a total of 74-h PW irradiation. Such a phenomenon cannot be explained on the basis of any thermal (temperature) effects and mechanisms [101]. It should be stressed that in all the experiments cited no aplastic bone marrow lesions similar to those described by *Yagi* after microwave exposures of rabbits [592] were seen. It may be concluded that the hematopoietic systems may be a convenient model for the study of low-dose repeated exposures for prolonged periods [20, 101].

It should be noted that a report of the induction of opposite effects in peripheral blood exists. *Deichmann et al.* [119] obtained an increase in RBC, granulocytopenia, and lymphopenia after single or repeated exposures up to 11 consecutive days of various duration (10 min to 7.5 h) of rats to 1.25-cm microwaves at 10 to 24 mW/cm^2.

The described changes in peripheral blood and the lymphoid system suggest that changes in the immunologic reactivity and the course of infections in irradiated animals are to be expected. Soviet authors found that the phagocytic activity of neutrophils decreases [244, 528, 586, 589], and resistance to infections is lowered. These investigations are discussed in detail by *Tchuchlovin* [542] in the monograph edited by *Petrov* [425]. These experiments are only in a preliminary stage. Modern concepts

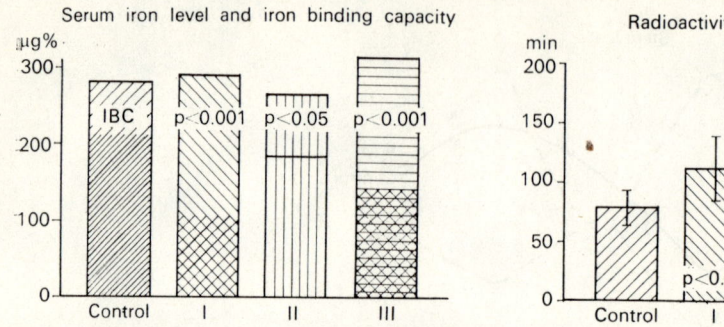

Fig. 60. Left — serum iron (level upper) part and latent serum iron-binding capacity in repeatedly irradiated rabbits. Right — plasma radioactivity half-time after ^{59}Fe injections. Control group: I, PW irradiation for a total of 74 h; II, CW irradiation for a total of 74 h; III, CW irradiation for a total of 158 h. Confidence coefficient p is indicated for the difference between control and experimental groups. From [495] and [102].

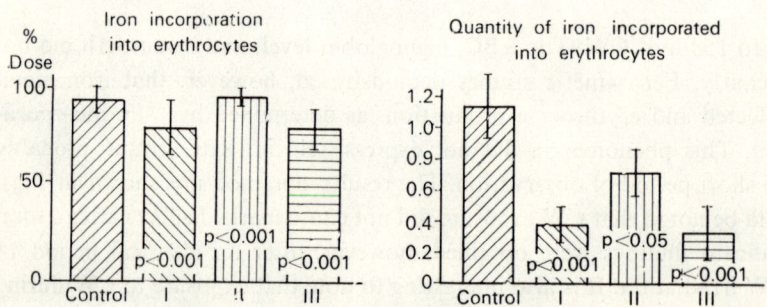

Fig. 61. Iron incorporation — left, percentage of dose, and right — mg/day/kg in same rabbits as in Fig. 60.

about the role of lymphocytes in immunologic responses make controlled studies on the immunologic reactivity of microwave-irradiated animals imperative, in view of the lymphoid system responses. Such a study is yet to be made. Finally, it should be mentioned that microwave radiation may affect blood coagulation [453].

The contents of this section may be summarized as follows:

1. Single exposures to high-power-density microwaves may cause various peripheral blood picture changes attributable to and explained by temperature increases and the resulting displacement of blood or of water in the body.

2. Low-dose repeated exposures in short-term experiments at or below 1 mW/cm^2 did not affect the peripheral blood picture to any significant degree.

3. In certain experimental conditions, especially after PW repeated or prolonged exposures, peripheral lymphocytosis, accompanied by stimulation of lymphopoiesis, may occur.

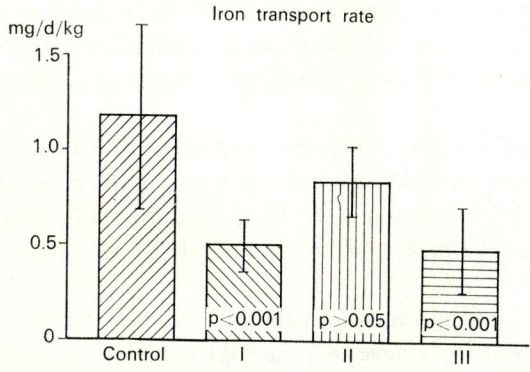

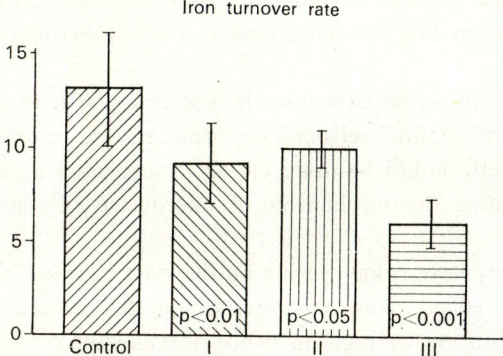

Fig. 62. Iron transport (top) and turnover rates in same rabbits as in Fig. 60.

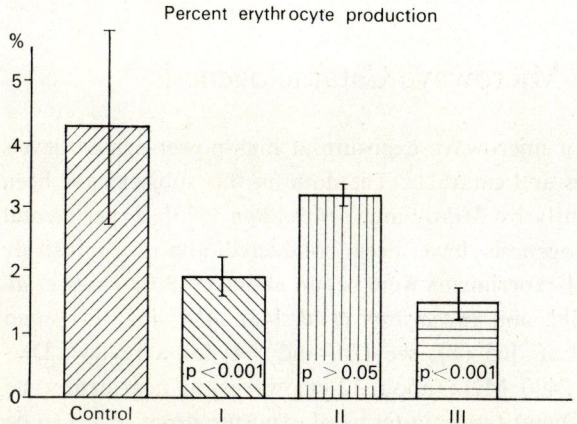

Fig. 63. Percentage erythrocyte production in same rabbits as in Fig. 60. Compare the effects of PW and CW irradiation in this and preceding figures. See the text.

4. The reaction of the lymphoid system merits special attention because of its possible bearing on the immunologic reactivity of the organism. Preliminary experiments have been carried out, and systematic studies employing modern immunologic methods are indicated.

5. Microwave irradiation (repeated low-dose exposures) may affect iron metabolism and red cell production; the mechanism of this effect merits special attention, as differences between PW and CW irradiation have been demonstrated. These differences point to an interaction at the molecular level that is difficult to explain by thermal effects mechanisms.

6. Microwave irradiation before ionizing radiation exposure may have beneficial effects on hematologic symptoms and the course of (ionizing) radiation sickness; this should be investigated further because of possible practical implications [361], as well as theoretical significance. Similar observations of a radio-protective effect on the second generation of bacteria descendant from microwave-irradiated organisms point to an involved mechanism for this phenomenon and the existence of genetic microwave effects.

7. Prolonged periods of repeated low-dose exposures induce changes in nuclear structure and mitotic abormalities in lymphoid cells and erythroblasts; chromosomal effects are a possibility. These effects should be more closely investigated because of their importance for understanding the mechanism of microwave interaction with living systems.

8. The phenomena mentioned may have some bearing on the eventual possibility of leukemogenic and carcinogenic effects; controlled studies on this subject are needed. Inbred strains of mice with a low or high incidence of leukemia could be used for such studies.

9. The blood-forming system seems to be a convenient model to study microwave effects both *in vitro* and *in vivo*.

Experimental Studies on Microwave Cataractogenesis

One of the best known effects of microwave exposure at high-power-density levels is the formation of lens opacities and cataracts. The data on this subject have been reviewed many times, most recently by *Milroy* and *Michaelson* [373]. Experimental studies on microwave cataractogenesis have been conducted almost exclusively by American authors. Controlled experiments were begun about 1948 by *Daily et al.* [108–111], *Duane* and *Hines* [138], and *Richardson et al.* [452, 454, 455, 457], and later developed by *Carpenter et al.* [65–68]; see [70] and [64] for a review. Diathermy exposures of 2450 and 2400 MHz (about 12-cm waves) at near-field zone were used in early experiments. Quantitation in terms of exposure doses seems to be doubtful. The results may be briefly summarized by the statement that exposure of eyes at high power densities causes either the appearance of immediate massive cataracts or delayed cataracts or opacities (with a varying latent period from a few to 14 days). Temperature increases were noted; nevertheless, in the early stages of these experiments metabolic effects were found and implicated as nonthermal mechanisms responsible for lenticular microwave-induced lesions. A decrease in pyrophosphatase and ATPase activities was reported by *Daily et al.* [111]. The influence of pathologic conditions, such as alloxan diabetes and pupillary diameter [457], was studied. Much discussion arose concerning the thermal or nonthermal effects involved. Temperature distribution was used as a guide for the determination of maximum absorption at various wavelengths in the eye. The results obtained with bovine eyes, which are similar to human eyes in size, are presented in Fig. 64. *Williams et al.* [570, 571] studied time and power thresholds for the production of lens opacities in rabbits exposed to 2450-MHz CW diathermy. The results obtained varied

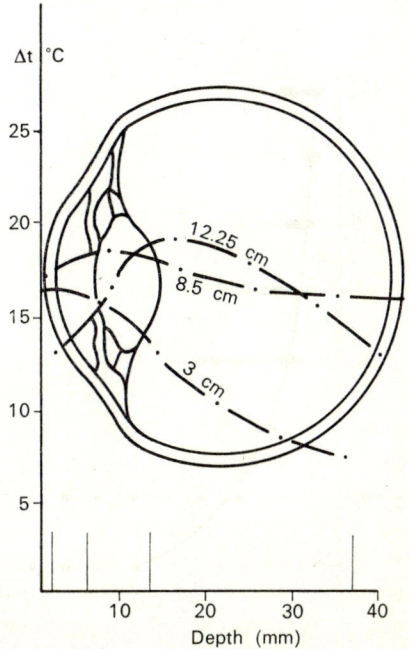

Fig. 64. Temperature rises in the eye at various depths depending on wavelength.

between 5 min at 0.59 W/cm² to 90 min at 0.29 W/cm². These authors stressed the similarity between microwave-induced and ionizing-radiation-induced opacities. *Cogan et al.* studied extensively radiation-induced cataracts [84–86], also noting these similarities. The available experimental evidence is evaluated differently by various authors. Two standpoints may be distinguished. One is represented by *Carpenter et al.* and was summarized in a recent review [64]. According to this author, the lens is highly susceptible to microwave radiation damage, at last in the microwave frequency range up to 10,000-MHz PW or CW. Cataracts may result from a single or repeated exposures. A latent period is characteristic for microwave cataracts; their morphology and latent periods may be compared to ionizing radiation effects. Biochemical changes occur, mainly change in ascorbic acid [350] and changes in DNA synthesis and mitosis in lenticular cells. Changes in membrane permeability occur, as evidenced by an increase in lens electrolytes and water. *Carpenter* [64] stresses the independence of lenticular opacities from temperature increases (critical temperature). The cataracts induced represent a permanent change in lens transparency. In view of this, the main points of *Carpenter's* views are the following:

1. Thermal and nonthermal mechanisms are operative in the induction of lens damage by microwaves.

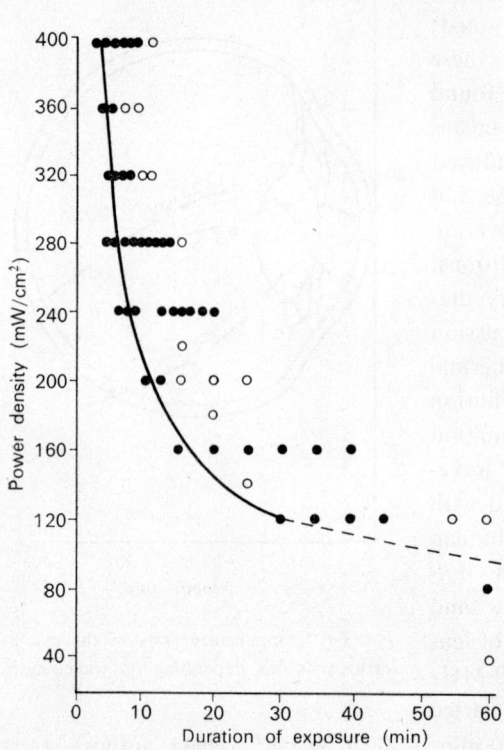

Fig. 65. Exposure duration power density threshold for lens opacities based on *Carpenter's* studies; • opacity, ○ no effect.

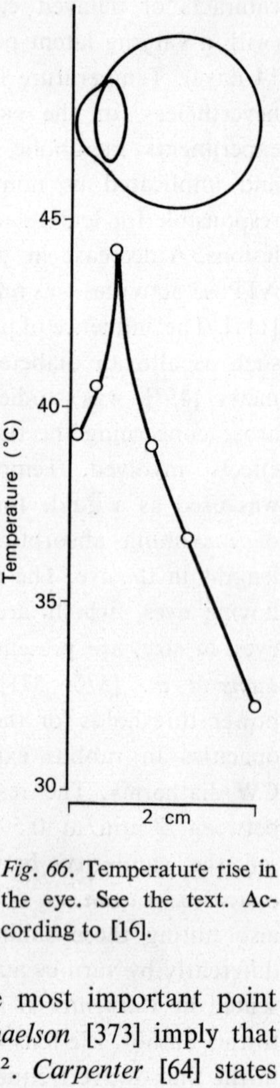

Fig. 66. Temperature rise in the eye. See the text. According to [16].

2. Cumulative effects of repeated exposures occur. The most important point is the threshold value for lens damage. *Milroy* and *Michaelson* [373] imply that this was determined by *Carpenter's* group as 80 mW/cm². *Carpenter* [64] states clearly that no opacities were observed after 1-h exposure at 80 mW/cm², but "the same dose, given daily for ten consecutive days, evoked opacity formation". In view of this statement, it is doubtful if 80 mW/cm² may be admitted as a threshold value for lenticular damage. In the present authors' opinion, such a statement should at least be qualified by adding "for a single exposure, which is not repeated for at least 48 h". Looking over the available data the authors found no basis for the determination of any power density level which is with certainty a threshold value for repeated, prolonged exposures, e.g., a few hours daily for several months. It may be added that *Kozuchowska*, *Tajchert*, and *Wojtkowiak* found changes in the electrophoretic mobility of lens proteins obtained from rabbits exposed for 4 weeks 2-h daily to 10-cm microwaves at 3 mW/cm² (in preparation for publication).

Opposite views are held by *Michaelson* [354]. Discussing the same experimental evidence, as well as personal experience in conjuction with the experiments of *Baillie* [15, 16] and *Birenbaum et al.* [46, 47], *Michaelson* concludes that the threshold limit for lenticular damage is about 100 mW/cm^2, and no evidence for cumulative or nonthermal effects exists. Figure 65 presents the time and threshold values for single exposures accepted by this group of authors, according to *Cleary* [80].

Birenbaum et al. [46, 47] examined the effects obtained by PW or CW exposure at various frequencies 0.8, 4.2, 5.2, 5.4, 6.3, and 70 GHz. No differences between PW or CW microwave effects could be demonstrated. The results agreed with the predicted depth of penetration, making possible the conclusion on that, depending on wavelength, posterior or anterior lens effects or corneal effects with increasing frequency are to be expected (compare Fig. 64). At 70 GHz, corneal effects predominate. This explains also signs of reaction of the conjuctiva reported by other authors.

Baillie [15, 16] examined microwave cataracts in dogs, using hypothermic animals. Two types of cataracts were distinguished: coagulative thermal, immediate effects and delayed. The latter are believed to be caused by disturbances in lens metabolism, initiated by temperature-dependent secondary effects. The investigations on temperature distribution showed greater temperature rises in the lens than in the vitreous humor. This may be a local heating phenomenon caused by reflections. These results are quoted by *Michaelson* as an argument that earlier temperature measurements, notably those of *Carpenter's* group, were unreliable. Figure 66 shows the temperature distribution in the eye of the dog as obtained by *Baillie* [15,16].

In view of the preceding data, it is impossible to take a definite stand. Further investigations are necessary to settle the controversy. One point must be firmly stated — sufficient data do not exist for the determination of threshold values for lenticular damage following prolonged periods of repeated exposures.

Miscellaneous Effects

Various isolated and usually contradictory reports exist on the influence of microwave exposure on this or that biological phenomenon. These observations do not contribute important data concerning the practical or theoretical aspects of microwave interaction with living systems. Most of the published papers are, however, cited in the monographs by *Presman* [438] and *Marha et al.* [336] or that edited by *Petrov* [425], and in certain instances in the review papers cited. On these ground numerous papers were omitted from this survey. The only exceptions are papers on the rather contradictory results of microwave heating of bone and its effect on bone growth [134, 154, 197, 240]. References to pertinent literature are to be found in these papers, which may be of certain interest to the physiotherapist or researcher. Another such series is the investigation of microwave effects on wound healing [277–279].

Comments on Experimental Studies on the Interaction of Microwave with Living Systems

The experiments cited in this chapter were conducted almost exclusively on mammals. Many papers on the interaction of radio-frequency radiation and magnetostatic or electrostatic fields with living systems based on experiments with fishes, plants, bacteria, and viruses have been published. These studies were omitted for several reasons. In most cases the frequencies used were outside the microwave range and, in consequence, outside the scope of this work. Most of the pertinent data and references can be found in *Presman's* monograph [438] or in the monograph edited by *Petrov* [425]; both books are available in English. Two excellent sources of data on the bioeffects of magnetic fields exist, one edited by *Barnothy* [33] and the second written by *Beischer* [38]. The interested reader is referred to these publications.

Comments on experimental findings were made in the final parts of each section. However, few more general remarks are necessary. The most important concern the problem of quantitation of the relationship between the exposure level and the bioeffect. In most papers insufficient or not wholly reliable data are presented. First, source characteristics are mostly incomplete. Irradiation conditions are not sufficiently described; the existence of reflections or scattering in the irradiation room is not mentioned. Not all experiments were conducted in anechoic chambers or with anechoic screens. In the present authors' opinion, the following data should be included in descriptions of experiments:

1. Source characteristics: type (if commercially available); frequency or wavelength (precisely or, if not, at least within a narrow band; such designations as decimeter or meter waves are insufficient) CW or PW; in the latter case, mean power and peak power (power density), pulse duration (pulse width), and repetition rate should be indicated; type of radiating element (or at least a general indication) waveguide exposures often necessitate a more detailed description of the arrangement and apparatus used.

In many instances military equipment is used, and in this case a certain reticence on the part of authors is understandable. Nevertheless, alterations are often possible that make a more precise description possible. The best solution would be specially designed equipment for experimental work; this is, however, very costly. In many instances the same equipment and arrangement are used in various biological experiments. In this case a special methodological paper containing a detailed description, as well certain data mentioned in the following items may be the best solution. Such papers should be published in easily accessible journals.

2. Irradiation conditions: field intensity (if PW, the mean and peak power), power density or electric and/or magnetic field intensities according to the experiment and equipment; irradiation in the near- or far-field zone or, given the characteristics of the radiating element, the distance from it; precautions against scattering and/or

reflections, anechoic chamber or screens; in many instances the expression "free space conditions" may be sufficiently descriptive; polarization; the orientation of the animal (object) in respect to the field and radiating equipment; the number of animals exposed simultaneously under the same conditions and their arrangement e.g., because of field distortion by the irradiated subject, irradiation of 4 or 12 rabbits simultaneously may not be readily comparable; detailed time table of irradiation regime; in certain instances with certain animals the time of the day or season may be important; are the animals restrained (how) or have they a certain freedom of movement (degree).

3. Some attempts at determination of the field distribution within (inside) the irradiated object should be made (if not sufficiently known, which is a rare instance); temperature measurements at various body points or preferably thermography as proposed by *Guy* may serve till better methods are developed.

4. Data on measuring equipment: the type if commercially available; reliability, sensitivity, method used, or appropriate references.

Some intimation in the text that the authors are aware of the complexity of quantifying the electromagnetic field is in certain instances desirable. Many of the points mentioned may be presented in a brief form by tables and diagrams; photographs may explain many problems.

It must be stressed that the instrumentation available for field measurements leaves much to be desired. Better equipment and methods are needed everywhere in the world. Another problem is the standardization both of equipment and methods. In the present authors' opinion, an international effort is needed to ensure comparability of data.

In view of the preceding, it seems that the biologist or physician must work in close collaboration with an electronics engineer and/or competent physicist.

Another general comment is that the controversy concerning thermal versus nonthermal effects does not contribute much to the understanding of the mechanisms of microwave interaction with living systems, if the biological methods used and interpretation of the biological results are inappropriate. It is difficult to gain insight into molecular and biochemical phenomena by using gross macroscopic methods (morphologic, physiologic) and making rectal body temperature measurements. Logical, step-by-step analysis of primary interaction and of the secondary effects — the chain of successive biological events — is necessary to advance knowledge in this controversial field. Both approaches to the primary events, the solutions stemming from *Debye's* theory and those based on quantum mechanics, are certainly legitimate. The quantum mechanical approach offers wider possibilities, which are till now unexploited. This is a subject for further carefully planned and executed investigations. At present it may be said only that several unclear points, well-documented phenomenological observations, await explanation:

1. Interference of microwaves and RF waves with the bioelectric function in nerve

cells and nervous system, muscle cells, and the heart (time relationship between the irradiation and phase of bioelectric function).

2. Behavioral effects and effects on conditioned reflexes; several explanations of the present experimental data could be advanced, all of a purely speculative nature.

3. Effects on mitosis, chromosomes, and the structure of cell nuclei; chromosome damage described as electromagnetic degeneration, human lymphocyte cultures, and the well established methods of analyzing the events following lymphocyte stimulation may prove useful for analysis of the interaction at the cellular level.

4. Effects in the hematopoietic system and lymphoid system; investigations on microwave effects on immunologic function may give interesting results.

It should be pointed out that effects on the cellular and subcellular level biomembranes and cellular organelles are largely unexplored. Biochemical, autoradiographic, and electron-microscopic methods may prove useful. Experiments on prolonged, low-dose, repeated exposure are needed; periods of exposure of several months should be investigated. The problems of genetic effects and eventual carcinogenic or leukemogenic effects should be investigated.

The questions raised on PW effects and their dependence on peak power should be clarified. Experiments carried out at various frequencies are needed; only selected bands have been investigated till now. The problem of intermittent irradiation and various irradiation cycle rats remains open.

Finally, it can be said that more thought should be devoted to the problem of the reciprocal relationship between the wavelength and the size of the irradiated object, as well as its external and internal geometry. Selection of animals for experimentation should take into account species-specific physiological aspects. This is even more important for the interpretation of results.

It should be stressed also that more investigation is needed on the effects of combined exposure to microwaves and other environmental physical factors. The same is true of simultaneous exposure to various microwave frequencies. At present it is difficult to decide how to evaluate the possible effects of simultaneous exposure to two or more microwave frequencies, each of a given power density. Should the power densities be simply added and is the effect proportional to the simple algebraic sum of power densities? If subtle nonthermal effects play an important role in the physiological response, it may be suspected that such simultaneous-exposure effects may be complex. The internal distribution of absorbed power within the body will vary depending upon the frequency, and a very complex picture will result. A biophysical and physiological analysis of such situations is urgently needed.

Health Status of Personnel Occupationally Exposed to Microwaves, Symptoms of Microwave Overexposure

CHAPTER 5

The overthelming majority of significant contributions in this field have been made by Soviet authors. Many translations and surveys of this literature have been published in English; regrettably a certain part of these publications are misleading, owing probably to insufficient familiarity with the language (Russian, English, or even both) and/or with the subject. A comprehensive survey of recent data is included in the monograph edited by *Petrov* [425]; the English translation of this book is competent. Excellent surveys were presented by *Healer* [219] and *Dodge* [127–131]. The reader is referred to these sources for additional data.

The main difficulty in evaluating any information about the health status of personnel professionally exposed to microwaves is the assessment of the relationship between exposure levels and observed effects. As often happens in clinical work, it is difficult to demonstrate a causal relationship between a disease and the influence of environmental factors, at least in individual cases. Large groups must be observed to obtain statistically significant epidemiological data. The problem of adequate control groups is controversial and hinges mostly on what one considers "adequate". All these questions must be discussed separately.

Precise quantitation of human exposure is possible only in the case of therapeutic or diagnostic applications of microwaves. Because of specific irradiation conditions and specifity of the material, the value of data obtained in this way is strictly limited. Controlled whole-body exposures of healthy adults at high-power-density levels or for prolonged periods at low power densities are inadmissible on ethical grounds, and such a possibility is categorically rejected by the present authors.

In view of the lack of adequate instrumentation, especially of personnel dosimeters, the quantitation of exposure during work is extremely doubtful. This is particularly the case where the personnel moves around in the course of their duties and are exposed to nonstationary fields, moving beam or antenna of radars for example, as well as to near- and far-field exposures alternately. It is impossible to evaluate within reasonable limits the exposure over a period of several years. Nevertheless, investigations of the health status of personnel exposed occupationally to microwaves demands examinations of large groups of workers exposed for various periods, if any statistically valid results are to be obtained. This difficulty is inherent to any medicoepidemiological studies, especially those concerned with exposure to various industrial or other environmental factors.

Perhaps the best solution out of these difficulties was proposed by *Gordon* [185]. On the basis of analysis of working conditions she divided the examined personnel into three groups:

1. Periodic exposure to high energy density levels, i. e., 0.1 to 10 mW/cm^2.
2. Periodic exposure to low energy density levels, i. e., 0.01 to 0.1 mW/cm^2.
3. Systematic exposure to low energy density levels.

The first group consisted of technical maintenance personnel and workers in repair shops and certain assembly factories. This group could be described briefly as workers in production, assembly technical maintenance, and repair of microwave equipment. It should be mentioned that a large part of this personnel was periodically exposed to fields of the near zone.

The second group consisted of technical maintenance personnel as well as ceratain categories of personnel engaged in the use of microwave apparatus, such as research workers.

The third group consisted of personnel engaged in the use of various microwave equipment, mainly at radar stations.

It should be pointed out that *Gordon* [180,182, 184-186] as well as other Soviet authors (e.g., [135, 136, 173, 207, 208, 273, 530, 551, 560]) analyze working conditions very carefully, taking such factors as air temperature and humidity, noise, and lighting into account. One factor that is usually omitted from these, as well as other, analyses is the problem of nonintended nonionizing radiation, in contradistinction to the ionizing radiation which is very carefully assessed.

In view of this it seems that the objections presented by *Dodge* [130] concerning the lack of pertinent data on the circumstances of irradiation and the possible influence of various other environmental factors in Soviet clinical papers are not altogether well founded. Attempts to present detailed data as to the source of microwave radiation, effective area of irradiation, position of the body in respect to the field, etc. (see *Dodge* [130]), for an individual worker for a period of several years would be misleading to an extreme degree. In the present authors' opinion, it is far better to present approximative evaluations than to create an impression of accuracy where none can be had.

It should also be pointed out that in most instances microwave workers are exposed to irradiation at various frequencies over the whole microwave range and often also to lower frequencies for instance in repair shops. *Gordon* [185] stresses the difficulties of differentiation between the effects of irradiation with microwaves and the effects of other environmental factors. The Moscow Institute of Industrial Hygiene and Occupational Diseases began systematic studies on the health status of personnel exposed to microwaves in 1948; a large series of clinical investigations was launched in 1953. The principal part of the clinical findings (up to 1966) was based on periodic examinations of "more than 1000" individuals observed for a period of over 10 years. One hundred of these workers systematically underwent special dynamic examinations along with a control group of 100 persons. The personnel examined had worked with microwave equipment for several years, in all cases more than 5. The age was up to 40 years, and according to *Gordon* the age factor did not influence the findings. The results of these observations were partly published by *Gordon* and her coworkers of the period 1953–1966 the clinical staff of the institute headed by the late *E. A. Drogitchyna* and consisting of *M. H. Sadtchikova, A. A. Orlova, S. F. Belova, V. K. Sokolov, N. A. Tchulina, I. A. Gelfan, S. N. Chmara, K. V. Glotova,* and *M. N. Smirnova*. These clinical experiences, as well as those of other Soviet authors, were analyzed in the monograph published in 1966 by *Gordon* [185]. This book is the most comprehensive source on the relationship between hygienic industrial factors, microwave irradiation, and the effects observed in microwave workers. The main conclusions reached by *Gordon* merit special attention in view of the many misunderstandings about the role of various environmental factors.

According to *Gordon* [185], the findings published by the Moscow group of clinicians concern personnel working under normal conditions of air temperature and humidity with low noise level. All the findings can be ascribed to the influence of microwave radiation.

Irrespective of exposure conditions (wavelength, intensity), a certain lability in the WBC may be observed; nevertheless, no typical (characteristic) changes in the peripheral blood picture may be found. Biochemical findings indicate certain changes in blood proteins and histamine content in serum and enzyme activity may be found; nevertheless, during a few days of rest, the findings become normal.

No conclusive data on the influence of microwave exposure on the eye are presented by *Gordon*, although an increased incidence of lens opacities and sporadic cases of cataracts are mentioned.

The chief attention in given to symptoms of functional disturbances in the central nervous and vegetative systems, as well as cardiovascular disturbances.

Centimeter waves influence the vegetative system as evidenced by vagotonia, bradycardia, and hypotonia. At high exposure levels (group 1), marked disturbances in cardiac rhythm, expressed by variability or pronounced bradycardia, were encountered. Marked and persistent hypotonia is characteristic. It should be mentioned that only in the centimeter-wave range were high-power-density levels encountered at working places. No experience with high-level exposure effects at other frequency ranges could be gathered.

Prolonged low-level exposure (group 3) to centimeter waves induces similar effects. The symptoms were less evident and easily reversible by periods of rest. It should also be pointed out that low-level exposure to millimeter waves is also characterized by cardiovascular effects, hypotonia and bradycardia.

Chronic exposure may lead to the development of a neurocirculatory syndrome characterized by three stages. In the initial stage, lability of blood pressure and cardiac rhythm are observed; the symptoms are easily reversible. In the second stage the symptoms become more accented; functional cardiovascular tests especially demonstrate the lability of the circulatory system. EEG changes indicative of disturbances in the diencephalon may be found. Thyroid function is disturbed; functional tests indicate hyperactivity [126]. A rather poorly defined "asthenic state" may develop. Subjective complaints concern headaches, excitability and irritability, fatigue disproportionate to effort, and pains around the cardiac region. The third stage is characterized by a more pronounced degree of expression of all these complaints and symptoms as well as ECG changes (prolongation of P deflection to 0.1 s and of the QRS complex to 0.1 s or more).

Exposure to decimeter or meter waves may induce EEG changes. Slow waves of low amplitude predominate; synchronized high voltage theta and delta waves are observed. According to *Gordon* [185] decimeter or meter waves affect the cerebral cortex and do not provoke such vegetative or cardiovascular symptoms as exposure to higher frequencies (centimeter or millimeter waves).

It should be pointed out that no differences were found between groups 2 and 3.

Gordon [185] draws attention also to the fact that radar operators, who are subjected to only very low doses of microwaves, if at all, may report complaints of eye fatigue and headaches, and demonstrate symptoms of hypotonia, bradycardia, and fatigue disproportionate to effort. In this case the symptoms or complaints are attributable rather to peculiar lighting, the necessity of paying attention to the radar screen, the posture during work, noise, or inadequate ventilation.

The observations on the health status of personnel exposed to microwaves are discussed in detail in the monograph edited by *Petrov* [425] by a group of Leningrad authors: *A. G. Subbota, E. V. Gembickii, I. G. Ramzen-Evdokinov, V. A. Sorokin,* and *V. G. Shiljaev*. This book was published in 1970 and the authors stress the fact that no objective symptoms or subjective complaints are encountered, if safe exposure limits are observed. However, overexposure may be met with, and in such cases or rather under such working conditions various complaints and symptoms may be found.

These authors distinguish a syndrome of acute microwave exposure and a syndrome of chronic microwave overexposure. Five cases of acute reactions were encountered. The syndrome is ill defined and seems to consist mainly of subjective complaints of headaches, nausea, vertigo, and sleep disturbances (sleeplessness). Objectively, hypertonia, changes in cardiac rhythm, and a skin rash may be encountered. EEG examination may show a decrease in amplitude of alpha waves. In all cases the symptoms were transient and after a few days of sedation and rest disappeared completely. No data concerning irradiation conditions were given.

It may be added that in personal experience (unpublished data, *Czerski*) two similar cases were seen. Both occurred in experienced technical personnel who realized that overexposure had occurred. In one case an engineer remained in the near-field zone of a high-power radar (10-cm band). Only the head was exposed for about 20 min at a (calculated) power density of about 50 to 70 mW/cm^2; the antenna was stationary. Realizing that overexposure had occurred, the engineer asked for an immediate medical examination in spite of a lack of any complaints. One hour after exposure the patient did not present any abnormalities, except signs of emotional unrest arising out of concern for his health. No abnormalities were found during 14 days of continuous observation (eye examination, blood and biochemical findings, physical neurological examination, EEG, ECG) and during follow-up studies 1, 3, 6, and 12 months later. The second case occurred also following overexposure in a complex field; the exposure level was approximately calculated ex post facts on the basis of local inspection of the work place and later measurements as in the preceding case. The wavelength was in the same range, about 10 cm; probably whole-body exposure to 30-70 mW/cm^2 during about 5 h occurred. The patient did not realize that overexposure occurred and suspecting an infectious disease, asked on the following day for medical advice because of a rash in the abdominal region and partially on the chest and back, vertigo, headache, and general malaise. All the symptoms disappeared on placebo medication during 4 days; no abnormal findings were obtained in spite of detailed clinical examination (medical, ophthalmologic, neurologic, laboratory, EEG, ECG) for 7 days after the exposure and on later (1 month, 6 months, 1 year) follow-up studies.

In the available literature only one report was found in which death was attributed to microwave overexposure [348a]. After exposure at the region of focus of a high-power radar, necrosis (thermal necrosis?) of the stomach wall and fatal hemorrhage ensued on the following day. Any causal relationship with microwave exposure was denied by *Ely* [148] 15 years later on the basis of records kept at the Army and Navy Institute of Pathology.

A careful analysis of the whole available literature (not only of the selected papers cited in the reference list), i.e., about 2000 papers, as well as of surveys and monographs up to September 1972 did not reveal any report or quotation of a report on injury following acute microwave exposure, microwave cataracts excepted.

The syndrome of chronic overexposure is described by the Leningrad group of authors [176, 207, 208, 276, 425] from two points of view. One concerns the individual variations in adaptation to the "microwave environment" [208, 425]. Subjective complaints of headaches, fatigue, and the like appear during the first 3 months of work, and reappear about the sixth to eighth month. After the first year, a period of adaptation of various duration occurs. A recurrence of complaints and symptoms of neurovegetative disturbances is to be expected after 5 years of work.

The chronic-overexposure syndrome [176, 276, 425] is characterized by subjective complaints consisting of headaches, irritability, sleep disturbances, weakness, decrease of sexual activity (libido), pains in the chest, and general ill-defined feelings of ill-being. On physical examination tremor of fingers and extended arms, acrocy-

Table 29

Frequency of hypotonia in personnel exposed to microwaves, according to Soviet authors; from [425], slightly modified.

Author and year	No. of individuals examined	Frequency of hypotonia %
Kerovski, A. A., 1948	87	38
Osipov, Ju. A., 1952	108	30
Shipkova, V. A., 1959	110	20
Orlova, A. A., 1960	525	26–33, various groups
Uspenskaja, I. V., 1961	100	30
Volfovskaja, R. N., et al., 1961	101	27–45, various groups
Frelova, L. T., 1963	172	25.6
Komarov, F. I., et al., 1963	53	22.6
Gembicki, E. V., 1966	210	14

anosis, hyperhydrosis, red or white dermographism, and hypotonia may be found. The frequency of hypotonia is shown in Table 29. As may be seen, the results vary greatly.

Examinations of the circulatory function included determinations of the velocity of propagation of the pulse wave. Various coefficients may be calculated and used for evaluation of vascular tonus and the state of the neurovegetative system. This method is widely used in the Soviet Union, but seldom elsewhere. Disturbances in the function of the circulatory system are demonstrable with the use of this method (see [425] for particulars). Except for signs of bradycardia, no significant findings are obtained by electro-, vecto-, and ballistocardiography. Mechanocardiography demonstrated normal or increased systolic and minute heart volume in individuals with hypotonia. On this basis, as well as from personal experience, *Gembicki* [176] questions the significance of blood-pressure decrease.

No characteristic changes in the peripheral blood picture are to be found. In about 25 percent of cases slight hyperactivity of the thyroid gland is to be found [126, 176].

The same group [425] discusses the problem of neurological findings, as reported by various authors. Many of the subjective complaints are attributable to the nervous system (Table 30).

As mentioned previously the authors are inclined to look upon these disturbances more as an expression of phasic adaptation or desadaptation to the working environment than as particular stages of a clinical syndrome. Examples of such clinical stages of various intensity may be found in the much-cited papers by *Klimkova--Deutschova* [270a, b]. Nevertheless, it seems more convenient to distinguish functio-

Table 30

Complaints of microwave workers (%) according to various authors; from [425], slightly modified.

Author and year	Headaches	Fatigue disproportional to effort	Sleep disturbances	Irritability	Abnormal sweating	No. of workers examined
Uspenskaja, N. V., 1963	37	31	29	9	7	100
Sadtchikova, M. N., 1963 (various groups)	12–39	20–35	—	8–27	—	447
Klimkova-Deutschova, 1963	43	39	35	—	—	73
Serel, 1959	43	4	45	10	—	103
Tjagin, N. V., 1966	33.5	46.2	25.3	9.6	25.5	573
Ramzen-Evdokimov and Sorokin, 1970	44	29	35	36	25	155
Controls						
Uspenskaja, 1963	15	22	2	10	—	100
Sadtchikova, 1963	8	10	—	8	—	100
Tjagin, 1966	10.8	5.9	8.7	—	2.7	184
Ramzen-Evdokimov and Sorokin, 1970	7	8	3	—	4	50

nal and emotional disturbances only occurring during the first period of work. Two first periods, one of early symptoms first year of work and a second of more expressed signs after 2 to 4 years, are usually not accompanied by objective findings. The third period, which may be seen in workers of over 5 to 10 years of exposure, is characterized by objective findings, disturbances in reflexes on neurological examination and EEG changes (a general slowing down and decrease in the amplitude of the bioelectric function). Emotional disturbances and neurovegetative dystonia may impair the ability to work, and a change of occupation is counselled. Signs of organic lesions in the nervous system are seen only very rarely in persons with over 10 years of exposure (fourth stage or period) and with frequent breaches of safety regulations.

It remains only to add that the Leningrad group looks somewhat sceptically upon the causal relationship of lens imperfections and microwave exposure ([425], p. 150 in the Russian text).

It should be stressed that the cited group of authors emphasizes that abnormal findings occur only in conditions of overexposure many times higher than accepted in the USSR safe exposure limits. The implication is that exposure levels were in such cases about 10 mW/cm^2 or more. The only exception is the recent report of *Kolesnik et al.* [274], who found a marked decrease in 17-CHS blood level response to ACTH injections in all of 35 microwave workers, as compared to a control group of 125 individuals.

As may be seen from Tables 29 and 30, groups of various sizes were examined. It is difficult to evaluate the statistical significance of all these data taken together. Taking all published reports at various times by different authors, at least several thousand microwave workers were examined. Nevertheless, because of gradual enforcement of safe exposure limits in the years 1959–1966 early Soviet reports cannot be compared with those published later (*Gordon*, 1972, personal communication). Changes in equipment and technology also may play an important role.

General conclusions are hardly possible. It seems that in normal conditions, if the USSR safe exposure limits are observed, no disturbances should be expected, except during the early period of work. If overexposure occurs, various transient functional disturbances may appear depending on the duration of overexposure over 5 to 10 years rather than the actual degree of overexposure, at least with present- -day technology. It may be added that precise data on exposure conditions are absent, as well as large series of investigations, excepting the experience of the Moscow Institute of Industrial Hygiene (*Gordon*, 1966, [185]) and that of the Leningrad group (*Petrov*, 1970, [425]). The most important Soviet publications on clinical aspects are listed in the subject index to the reference list (see the entries on clinical symptoms and health status of personnel exposed to microwaves).

Only a few papers were published by authors from other countries. The early American reports of *Daily* [106] and *Lidman* and *Cohn* [308] cannot be applied to present-day conditions. *Barron et al.* [34] did not find and significant devations from normal; this report was singularly uniformative on exposure conditions. The few Czechoslovakian reports (*Klimkowa-Deutschova* and *Sercl*) agree with Russian authors. Similarly, *Miro* [387] also confirms the occurrence of nervous system disturbances and asthenia, as described by Soviet authors, in a group of 36 French microwave workers.

A large series of investigations was carried out in Poland, mainly at the State Institute of Industrial Hygiene by *Minecki* [374, 375, 377, 378, 380] and at the Warsaw Institute of Aviation Medicine [20, 24, 44, 101, 103, 104, 215, 217, 335, 403a] as well as by other groups (see the entries clinical symptoms and health status in the subject index to the reference list). All these findings will be discussed jointly.

Safe exposure limits and regulations enforcing various safety measures and precautions were introduced in Poland in 1961 (see Chapter 6). As in other countries, only gradual enforcement was possible. In view of this, all examinations carried out before 1962 concerned persons subjected to uncontrolled exposure. The possible exposure levels could be evaluated only *ex post facto*, and only approximately. In the period 1962–1968 data on exposure levels based on power density measurements and the analysis of working conditions became available; most examined individuals had, however, a shorter or longer uncontrolled-exposure history. Only after 1965 were health requirements for persons working in microwave-irradiated environments [546] really exact. At the same time a change of occupation was counselled for many microwave workers because of various health defects, not necessarilly related to microwave exposure, which could be aggravated by further work. At the same time a new "generation" of microwave personnel undertook work. These individuals passed through an initial examination (before commencement of work) and were examined periodically during work in approximately known and controlled conditions. In view of this, the publications concerning the health status of personnel professionally exposed to microwaves may be divided into papers concerning.

1. Persons having a history of longer or shorter periods of work under uncontrolled conditions, with exposure levels undetermined or calculated (or rather guessed at) *ex post facto*.

2. Persons with a history as in item 1 and a period of work in a controlled environment.

3. Persons examined before work was undertaken and working in controlled environments.

Numerous Polish papers concern the first group [44, 202, 203, 214, 215, 283, 285, 374, 376-378] or the second and third groups jointly [20, 24, 101, 104, 124, 140, 143, 235, 236, 250, 325, 335, 339, 409a]; only a very few deal exclusively with the third group [235, 602-604]. All these findings, as well as additional unpublished data, served as a basis for the determination of the new Polish safe exposure levels [104], as described in Chapter 6.

Selected groups of microwave workers were first examined at intervals of 3 months, later 6 months, and finally each year. All the persons examined were usually divided into three groups (see [24] or [409a]) for examples:

1. "E"*: low-level exposure in the order of tens of microwatts per square centimeter, usually in far-field conditions or in complex fields in closed rooms where only very low power equipment was installed.

2. "ES": mean-level exposures in far- and near-field zones, where no important nonintended radiation was expected; the measured and expected levels were in the order of hundreds of microwatts per square centimeter up to about 1 mW/cm^2.

3. "R": high-level exposures in the order of 1 mW/cm^2 up to 10 mW/cm^2, and in certain instances even more; or exposures corresponding to group 2, but with a frequent exposure to nonintended radiation expected.

No attempt was made to differentiate between exposures at various microwave frequencies, as most individuals available were exposed at this or other times over the whole microwave range. No adequate control group could be found because of difficulties in finding individuals working in sufficiently similar conditions (temperature, noise, humidity, time of day, etc.), with sufficiently similar economic and social position, with similar everyday living habits and conditions, and belonging

---

* The letters correspond to Polish designations of these groups in the original papers.

to the same age group. It was decided to analyze the material according to the total period at work within the group and according to the exposure level (among groups). When making the last type of analysis, it was possible to collect groups that were reasonably similar in all respects (socioeconomical and psychological included) save the level of exposure.

The examinations were conducted partly by the authors of the publications listed and their collaborators mainly the team of the Institute of Aviation Medicine in Warsaw on selected groups of several hundred individuals (see references) and partly at various Polish Health Service posts, according to a detailed enquiry sheet. These latter examinations were verified at random, compared to personal results, examined for contradictions, and evaluated for reliability (the enquiry sheet was specially constructed, and contained many questions to which the answers were mutually exclusive). The examination consisted of a general physical checkup, detailed neurological and ophthalmological examinations, peripheral blood examination, urinalysis, and chest X-ray. Case histories concerned individual, family and professional history exposures. ECG and EEG were made in selected cases, consisting of a total of several hundred persons (see [143, 339, 409a], respectively). All cases where pathology related to any known factor was found, were excluded from the analysis, which finally was based on several thousands of "healthy persons exposed to microwave environment." This analysis is being continued with respect to the third group; successsive examinations of the same individual will be compared to determine the eventual dependence of changes on the total period of work in different exposure conditions, starting from a detailed examination at the beginning of work. It should be stressed that all persons examined passed a physical checkup and a more or less detailed medical examination before work was started; in many instances in the first and second groups* the results of these examinations were either unavailable or considered insufficient or unreliable. It must also be mentioned that radar operators, who are working in particular conditions and are not exposed in any significant degree (periodic exposures on the way to work and back), were not included in this analysis.

The results confirm the previously cited findings of Soviet authors on the periodicity of the occurrence of subjective complaints. These were met only in the mean- and high-level exposure groups. It must be stressed that the low-level exposure group did not show any objective symptoms up to 10 years of work (no persons with a longer period of exposure were found in this group). It was decided to use this group as a reference standard. It seems that this solves best the problem of control groups.

Subjective complaints were exactly the same as described by Soviet authors (see Table 30). The presence of such complaints is characteristic for up to 70 percent of the persons subjected to uncontrolled exposures (early findings). Headaches

---

* Controlled or uncontrolled exposure conditions.

and fatigue disproportionate to effort occur in 47 and 45 percent in the "R" group, in 30 and 34 percent in the "ES" group, and 30 and 30 percent in the "E" group during the first year of work; they disappear for 2 years, recur during the period 3 to 5 years of work, and may reappear in certain individuals after 5 to 10 years. Abnormal excessive sweating during the night has a similar time dependence and was found in 68.8 percent in the "R" group, 33.4 percent in "ES", and 22.5 percent in "E" during the first year of work; in the period from 3 to 5 years of work, the respective values were 14.5, 15, and 7 percent. Later this symptom was not observed. Changes in blood pressure occurred only in the "R" group, the percentages of hypotonia being 18 during the first year, 14 in the period from 1 to 3 years, 6 in the period from 3 to 5 years, 8 in the period from 5 to 10 years, and 11 over 10 years of work. In the remaining groups this percentage was less than 1. It should be added that no correlation with changes in heart-beat rate could be demonstrated. No changes in ECG were found. One finding that shows definite correlation with the duration of work and exposure level is loss of hair.

The peripheral blood picture shows variations within the normal distribution range in the "E" and "ES" groups. In the "R" group diversified WBC responses were found during the first year of work. After 10 years of work 10.5 percent of workers in the "R" group show absolute lymphocytosis usually accompanied by monocytosis, the total WBC being over $10,000/mm^2$.

Neurological examinations are difficult to evaluate. Many of the physicians who carried out these examinations differed in the evaluation of reflexes, dermographism, signs of irritability, and the like. In view of this, the only means of obtaining objective results were EEG studies. These do not demonstrate any abnormalities in the "E" group. In the "ES" and "R" groups, depending on duration of work and degree of exposure, a definite decrease in the number and amplitude of alpha waves occurs. Theta and delta waves and spike discharges may occur. The response to photostimulation is decreased. The most impressive finding is poorly expressed bioelectric activity after more than 5 and even more so, 10 years of work. These findings are in accord with the results obtained by Soviet authors. It should be noted that *Pivovarov* [432] reported that chronic microwave exposure increases the threshold of stimulation of the senses (vision, auditory function, skin receptors) in men.

It should be stressed that a rather specific phenomenon occurs in microwave workers [143]. Intravenous administration of cardiazole (Metrazol) may be used for provocation of discharges (preconvulsive discharges) in the EEG, convulsions, or shock. According to the literature, this phenomenon is dose dependent and a cardiazole threshold exists; doses of 7 mg/kg body weight have no effects. Intravenous administration of 500 mg of cardiazole in 10 ml of saline (1 ml/30 s with 30-s intervals) does not provoke any effects in a normal adult male. In microwave workers with over 3 years of exposure, theta waves, theta discharges, spike discharges, and even convulsions occurred. Twelve persons were examined; in eight the test could not be completed. The study was discontinued, as the test was considered dangerous for the patient. It should be pointed out that this phenomenon was studied extensively in rabbits [30], and a decrease of cardiazole tolerance in irradiated animals may be considered as established.

Certain authors [367]; see also [425] have attempted to determine a microwave overexposure clinical syndrome. *Czerski et al.* [103] have published a paper describing a case of such syndrome, where psychoneurological symptoms predominated. Looking back, most statements contained in this paper should be reevaluated. According to the current opinion of the present authors, the existence of a well--defined nosological entity, similar to the "vibration syndrome", of a "microwave overexposure syndrome" remains to be demonstrated. Nevertheless, such a possibility cannot be excluded and further analysis is necessary. At present it must be said that in certain instances syndromes of neurological disturbances (without organic lesions) and signs of neurosis, accompanied by a poorly expressed bioelectric function of the brain, are found in microwave workers following long periods of exposure. These patients may be incapacitated for further work and even normal everyday life. A causal relationship to microwave exposure seems to be probable. It is, however, difficult to prove satisfactorily. If lens opacities and/or absolute lymphocytosis, granulocytopenia, and slight anemia are also present, this suspicion is strengthened.

In the course of work connected with health surveillance and risk analysis, the present authors encountered "clusters" of certain deviations from normal in factories or other working places where exposure levels were exceptionally high, i.e., about 10 mW/cm^2 or more during about 1 h/day. In such places exposure to nonintended radiation could also be expected. The abnormalities consisted in the presence of 0.5 to 2 percent of otherwise healthy persons with deep bradycardia (less than 50/min) and signs of impairment of heart conductivity in the ECG, and various percentages of workers with stomach ulcers or peripheral blood picture changes (slight anemia, lymphocytosis or granulocytopenia, or persisting unexplained granulocytosis). Such groups were usually too small to draw any valid conclusions, so only an impression remains that working conditions had "something to do" with these phenomena. This impression is strengthened by the fact that after introduction of rigorous health and exposure surveillance, as well as a partial exchange of personnel, no further cases were noted.

Preliminary experiences ([275, 336]; show that carefully selected personnel (i.e., without any health defects initially) exposed to field intensities corresponding to new Polish safe exposure limits [401] or even slightly higher values, were in a more satisfactory state of health than a group of personnel unexposed to microwaves. These comparisons were made by pairing individuals for age, socioeconomic conditions, and working conditions other than microwave exposure. It should be stressed that the microwave-exposed group was a highly preselected one; i.e., all the individuals were subjected to initial medical examinations and only perfectly fit men were allowed to work in microwave-exposure conditions. The control group was inadequate in this respect. No initial selection was made according to health criteria. These observations concern small groups, observed for 3 to 4 years, and no statistically valid conclusions can be drawn. These facts are mentioned as a suggestion for further approaches to the problem of the health status of personnel exposed to microwaves rather as conclusive data.

A problem that demands further investigation is the influence of microwave irradiation on the auditory organ. Impairment of auditory acuity during irradiation [275, 432] has been reported. *Frey* [163, 168] described the hearing of certain frequencies by man. The mechanism of this phenomenon was much discussed and it may be said that it remains unclear till now. An explanation of this observation is important for considerations of the mechanism of the interaction of microwaves with the nervous system; the practical significance may be questioned.

Another controversial problem is the occurrence of lenticular opacities and microwave cataractogenesis in microwave workers. Experimental evidence (see Chapter 5) proves definitely that microwave irradiation may produce cataracts; the threshold is about 100 mW/cm^2 for single exposures and about 80 mW/cm^2 or less for repeated exposures. Frequency plays an important role by determining the depth of penetration and maximum energy absorption. The most effective frequencies seem to be in the 1000- to 3000-MHz range, although no sufficient systematic research has been done at higher frequencies. Beginning from 30,000 MHz, corneal and conjuctival damage is to be expected. No clear evidence exists on the influence of PW irradiation, especially at very high pulse peak powers.

Bilateral or unilateral cataracts in microwave workers were described. In 1969 *Zaret et al.* [594] found 44 such cases, 42 personal observations and those of *Hirsch* and *Parker* [229] and *Shimchovitch* and *Shilayev* [534]. The total number of reported cases may be estimated at over 50 (see [31, 45, 216, 250, 255, 292a, 433, 595]). The available evidence indicates that inadvertent or simply lightminded, careless exposure of the eye to high power densities (looking into a wave guide or a window slit) may induce cataracts. Cataractogenic exposures should not occur in the course of normal work, at least with present-day technology.

The occurrence of lenticular imperfections was studied in 736 radar workers and 559 control individuals by *Zaret et al.* [594]. On examination using a slit lamp, these authors have distinguished the following:

1. Minute defects: granules and vacuoles.
2. Opacifications: irregular cloudy opacities or strations.
3. Relucence: opalescence and light points.
4. Zonular defects: thickenings or striations.
5. Posterior polar defects: opacities which may be of diffirent sizes, in certain cases forming an incipient or developed polar subcapsular cataract.

These changes were classified as insignificant as to number and/or intensity 0 degree, slight (1°), moderate (2°), or (3°). It was demonstrated that a difference in opacification between the radar workers and control groups exists at the 2.5 percent confidence level and in posterior polar defects at P less than 0.0005 level. The authors conclude that the lens may serve as an individual dosimeter for cumulative microwave injury.

Belova [45], on the basis of the examination of 370 microwave workers, and *Majewska* [325], on the basis of the examination of 200 microwave workers and 200 control persons, are inclined to admit the causal relationship between microwave exposure and lens opacities. However, a coauthor of one of the first reports on microwave cataracts, *Shilayev* (in [425]), seems to be sceptical on this point, as mentioned previously. The studies of lenticular imperfections carried out at the Institute of Aviation Medicine in Warsaw by *Janiszewski* and *Szymańczyk* [250] point to a close correlation among length and level of professional exposure (classified as above), and the incidence of lenticular imperfections. Perhaps the most conclusive studies are those of *Żydecki* [602- 604]. Using a slit lamp after dilatation of the pupil, this author distinguishes five grades of lenticular transparency:

1°. Complete transparency; no opacities are demonstrable.

2°. Single small opacities; dust-like or point-like or radial strations, which may be counted and do not influence visual acuity.

3°. Similar to grade 2°, but numerous; no tendency to increase in number or size; no visual acuity impairment.

4°. Similar to grade 3°, but with a tendency to increase in number or size on successive examinations.

5°. Any lenticular imperfection causing impairment of visual acuity.

The author admits that all five grades may be inborn, caused by various injuries to the eye, or eventually related to microwave exposure. To test this last possibility, 3000 persons were examined; 1000 were microwave workers of which 542 were technical personnel (group A_1) working in the near-field zone and exposed to nonintended or dispersed and reflected radiation. The author states that in principle, the 100 and 1000 $\mu W/cm^2$ safe exposure limits were observed (see Chapter 6), but admits the possibility of much higher nonintended exposures. The personal estimate of the present authors would tend to set the exposure levels higher, at from several up to 10 mW/cm^2 during about 4 h/day on the average. Exposures over 10 mW/cm^2 for significant periods of a few minutes could occur sporadically; this is, however, not very probable in view of the safeguards and precautions adopted among this group of personnel. A second group of 458 microwave workers (group A_2) were exposed to far-field zone conditions, according to the author, "in principle", at about 10 $\mu W/cm^2$ (our estimate being from tens up to a few hundreds of microwatts, but certainly below 1 mW/cm^2). The age of the microwave workers varied between 18 and 60 years. A control group of the same age (group B), profesionally unexposed to microwaves or any other radiant energy and selected for similar living and working conditions, consisted of 1000 individuals. A second control group of 1000 kindergarden and school-children from the age of 5 to 17 years was examined (group C) with the aim of determining the incidence of inborn lenticular defects. The author points out that the microwave workers are a preselected group because of the health requirements for admission to work. In control group C, grade 1° transparency was found in 86.8 percent in young children and in 60.1 percent in older, 68.8 percent on the mean. Grades 4° and 5° were seen in seven cases (three and four respectively). The conclusion is that the percentage of grade 3° increases with age. This is confirmed by the results obtained in group B, where among younger adults (up to 25 years) 43 percent were classified as grade 1°, but only 6 percent of those persons over 50 years of age were so classified. Analysis of regression curves for both control groups demonstrates that the frequencies of grades 1° and 2° decrease and those of grades 3°, 4°, and 5° increase. Overall mean percentages for group C are 29.7, 50.3, 16.6, 2.6, and 0.8-percent for grades 1° to 5° respectively. For group A_1 the figures are 13.0, 62.0, 19.7, 4.8, and 0.5 percent for grades 1° to 5°, respectively. Group A_2 did not differ significantly from group B. An analysis of age groups and duration of exposure periods (group A_1, A_2, and B) led to the conclusion that exposure levels affect the incidence of lenticular opacities and/or lenticular defects; the apparent influence of duration of professional exposure in group A_2 is equal to the dynamics of age changes. It must be stressed that the statistical treatment of data is extremely careful and does not leave room for doubts. These data may be interpreted, as is done by *Michaelson* [354] with the findings of other authors, to mean that microwave exposure accelerates the aging process of the lens. It may be concluded that eye protection is necessary for microwave workers, at least in certain frequency ranges, and that for prolonged periods of occupational exposure a safe limit of 1 mW/cm² or less should be observed (see Chapter 6).

An evaluation of microwave effects on the eye is complicated by the reports of *Tengroth et al.* [547] and *Marner* [338] on retinal lesions in 17 of 50 Swedish microwave workers. This observation caused the lowering of the safe exposure limit from 10 to 5 mW/cm², at least in the factory where these observations were made [337, 338]. It may be recalled that *Belova* [45] and *Ju-Tchzhin* [255] earlier had drawn attention to retinal lesions induced by radiant energy exposure. It is to be noted as a significant fact that discussions with our colleagues, the ophthalmologists examining microwave workers, disclosed that retinal changes were noted and not reported, as "having no relation to microwave exposure". This point should certainly be clarified by examinations of larger groups of microwave workers.

A separate problem, which is only incidentally related to the subject of this survey, is ocular fatigue and its consequences noted among radar operators [41, 61, 185, 459, 602]. This depends on lighting, type of screen, distance from it, and similar factors [185, 268, 385, 386]; in the opinion of the present authors, microwave exposure cannot be implicated as a contributory factor.

The survey of the literature cited allows several comments; the main points are the following:

1. Uncontrolled professional exposure to microwaves leads to the appearance of vegetative and central nervous system disturbances, asthenic syndromes, and

such like chronic prolonged exposure effects, which are well documented by early Soviet, Polish, and Czechoslovakian reports; the pathogenesis of these syndromes may be controversial (see [425]), but their existence cannot be denied. Similar observations were made by *Miro* [387] in France, and in the United Kingdom and the United States according to a personal communication made by *Mumford* to *Seth* and *Michaelson* [490]. The negative reports found in "Western" literature consist of a study by *Barron* and *Baraff* [34] and *Martner* [338]. In both, data on exposure and on working conditions of the personnel examined are insufficient; the exposure seems to be controlled within the U. S. safe exposure limits. The medical examinations were restricted by eye examinations, as well as general routine physical and medical examinations certain laboratory investigations, such as chest X-ray. No neurological findings, EEG, or ECG data are presented. The American report dates back to 1958. It is interesting to note that a high incidence of eye pathology was reported. Many unsupported statements on the lack of effects of occupational exposure to microwaves may be compared with the many papers on hazard evaluation, risk prevention, and technical precautions [51, 53, 225, 226, 228, 462], in several instances published by the same authors who question the reliability of clinical observations.

2. Controlled occupational exposure of healthy adults seems to have no untoward effects if the new Polish safe exposure limits [461] or even more so the conservative Soviet ones [591] are observed. Exposure for prolonged periods (over 5 or 10 years of work) at power density levels higher than these limits leads to nervous system disturbances, as evidenced by discrete symptoms on neurological examination and EEG changes [335, 339] consisting of a poorly expressed bioelectrical function.

3. Various phases in adaptation to the "microwave environment" (the initial complaints and neurological signs and hematological changes characterized by diversified objective findings during the first year, and their later recurrences), as described by *Subbota* [510, 517]; also in [425], were described independently by Polish authors and in the present authors' opinion are well documented.

4. Lenticular opacities induced in microwave workers by occupational exposure are sufficiently well documented to limit exposure over 1 mW/cm^2 (at least for certain frequencies) and to require special precautions. Microwave cataractogenesis at higher power densities is a fact, irrespective of its "thermal" or "nonthermal" mechanism. The influence of microwave exposure on the lens with inborn or age defects remains to be investigated.

5. Acute exposure at high levels (over 10 mW/cm^2, and even more so at 100 mW/cm^2) may be expected, depending on the duration and exposure conditions, to induce various local or generalized thermal injuries. Depending on localization and temperature increase, various extremely diversified effects may be expected. Thermal necrosis of a vital organ or its part may lead to death. The case reported by *McLaughlin* [348a] of microwave death may not be proved, but such or similar cases can be envisaged and certainly may occur, given sufficiently unfortunate con-

ditions of exposure in the vicinity of high-power radars. Thermal stimulation (metabolic effects) of limited body areas (nervous system, endocrine glands, liver) may induce, if repeated, various delayed or secondary effects. The influence of repeated thermal stress induced by discrete, frequent whole-body temperature elevations is unknown. The influence of repeated exposures on biological rhythms and their synchronization (of rhythmic activity changes of various organs in case of limited--body-area exposure) should be explored [569].

6. The last point, which is most important and cannot be sufficiently emphasized, is that all available data concern healthy human adult exposure, mostly men. The effects of intermittent or continuous exposure of children living near radar installations or television transmitters is completely unexplored. Children may be expected, because of body size and geometry, to absorb microwave energy differently from adults. Exposures of 4 to 8 min/day to microwave irradiation at low mean (tens or hundreds of microwatts per square centimeter) and very high peak power densities are sufficiently real for children as to cause concern. One of the present authors was impressed by the sight of children playing with low-power radars in a yachting harbor, full of seagoing pleasure craft. On the present evidence nothing can be predicted within any reasonable limits of probability. Investigations on the problems of the eventual effects of microwaves and even more so lower-frequency nonionizing radiation, are urgently needed. For proof of the urgency of this need, the American reader is referred to the many U.S. publications [51, 80, 348, 462], especially by *Mills et al.* [371].

CHAPTER 6

Safe Exposure Limits and Prevention of Health Hazards

In Chapter 5, it was shown that the bioeffects of microwaves and the mechanism of interaction with living systems are insufficiently known. Nevertheless, because of practical, actual exigencies of modern life, human exposure must be limited and possible microwave health hazards prevented. In view of the many uncertainties and differences of opinions, it is not surprising that the safe exposure limits proposed in various countries differ greatly.

Michaelson [351, 354], discussing human exposure to nonionizing radiant energy, points out that in all safety standards, rules, and laws the principle of calculated risks is admitted in a more or less open manner. Quoting discussions on ionizing radiation protection, the author stresses that many human activities involve a certain element of risk, and that often decisions are made without formal evaluation of probable gain or loss. The statement that the progress of science and technical advancement involves many potential hazards is an obvious truism. Nobody would propose to stop progress. Nevertheless, the necessity of directing and controlling this process so as to prevent any untoward effects seems to be no less obvious. Laws and recommendations are introduced with the aim of minimizing unfortunate consequences. Better designs are developed. Many of the actual existing solutions, laws, or recommendations are only temporary. It is also obvious that economic and social conditions, the balance of public gain or loss, influence legislation.

Because of the professional competence (or incompetence) of the present authors, only the biomedical aspects of the existing radiation protection guides and/or laws will be discussed; i.e., the published reasons for accepting a given safe exposure limit will be cited. From the many voices in the discussion, only those will be cited which concern meritorious biomedical aspects of this problem.

According to *Michaelson* [354], the first proposal to limit human exposure to 10 mW/cm^2 was made by *Schwan* in 1953 on the basis of simple physiological premises. *Schwan's* reasoning was presented in detail in two comprehensive publications dating back to 1956 [480, 481] and confirmed by this author in 1968 [476], 1969 [471], and 1971 [473]. Analysis of microwave energy absorption in tissues (see Chapter 3) led *Schwan* to the conclusion that the effective cross section of man may be assumed to be equal to 1.0 (incident energy per unit area is equal to the absorbed energy), allowing an error of 50 percent. Heat-balance characteristics of standard man in standard conditions permit the acceptance of 10 mW/cm² as a safe limit, because no temperature increase is to be expected, even during infinite exposure [480, 481]. *Schwan* believes [476] that injury to the eye, damage to the blood--forming system, or "other physiological disturbances" may occur following exposure at power densities within the 50 to 100-mW/cm² range or more. The author [473] confirms the impossibility of microwave interaction with nervous system cells (see Chapter 3). Stressing the many uncertainties concerning the biological effects of microwaves, especially in such effects as microwave-induced chromosome aberrations, *Schwan* [476] concludes*: "Many data must be collected before all questions connected with microwave hazards will be solved. The existing data are nevertheless sufficient to determine the safe exposure level ... for prolonged exposure it is equal to 10 mW/cm² ... this value may be increased for short term exposures."

Schwan [473, 474, 476] cites selected references on conditioned-reflex disturbances, cardiovascular effects, and clinical observations [135, 184, 294, 314, 440], and maintains that no satisfactory reasons exist for lowering the safe exposure limit. It may be concluded that *Schwan* agrees fully with the U.S. Army regulations [87] and U.S.A. Standards Institute recommendations [468]. The text of these and comments made by certain authors [416, 354] allow one to suppose that both recommendations are based on *Schwan's* views, which may be briefly presented probably without gross misinterpretations as follows:

1. The principal bioeffect of microwaves is temperature increase in the irradiated object (thermal effect).

2. Heat-balance characteristics of man allow infinite exposure at 10 mW/cm²; higher values may be accepted for short-term exposures.

3. Cataract or lenticular opacity formation are to be expected at power densities below 100 mW/cm²; repeated exposures at 80 mW/cm² may, however, induce such effects. Functional disturbances may occur following repeated exposures at 50 mW/cm².

4. Biophysical considerations permit one to exclude the possibility of microwave interaction with nerve cells.

5. No sufficient evidence concerning the possibility of induction of untoward effects of microwave irradiation in man at power densities below 10 mW/cm² has been presented.

The actual U.S. Army Standard dated 1965 [87] (see also [416]) replaced the regulations from 1958 [561] and is in effect till now (April 1975). Unlimited exposure below 10 mW/cm² is allowed; exposure at power densities higher than 100mW/ /cm² is considered dangerous. Within the range from 10 to 100 mW/cm², exposure for a limited time is allowed, according to the formula

$$t = \frac{6000}{W^2}$$

---

* This is a retranslation from the Russian text, not a verbatim quotation.

where t = exposure time,

W = mean power density in milliwatts per square centimeter.

Figure 67 illustrates these rules. In addition, it is recommended, whenever possible, that the exposure level around and inside military installations be minimized. Additional recommendations are as follows [146]: 0.01-W/cm² (10-mW/cm²) power density in the radio-frequency range should be considered hazardous; appropriate warning signs must be placed in zones where such power densities occur. Intended human exposure at power densities over 0.01 W/cm² should be as a rule avoided; exposure to power densities of 0.01 to 0.1 W/cm² are allowed with due precautions. Attention is drawn to incidental ionizing radiation hazards in the vicinity of equipment operating at voltages over 18 kV.

The recommendations drawn up by the C 95.1 committee of the U.S.A. Standards Institute (USASI-C 95.1) do not determine the upper limit of permissible exposure [468, 491]. This limit is indirectly defined by the recommendation:

In normal environmental conditions the density of electromagnetic energy flux at frequencies between 10 MHz to 100 GHz should be 10 mW/cm² on the mean, as averaged over any 0.1-h 6-min period. This means that power density of 10 mW/cm² for a 0.1-h period and 1 mW/h/cm² over any 0.1 h are considered a safe exposure level. No distinction between PW or CW exposure is made.

From the practical point of view, U.S. Army and USASI recommendations do not differ significantly. As exposures below 2 min are difficult to control in practice, the practical U.S. Army limit is 55 mW/cm². USASI recommendations allow 1-min exposure at 60 mW/cm². It should be noted, however, that this exposure must be

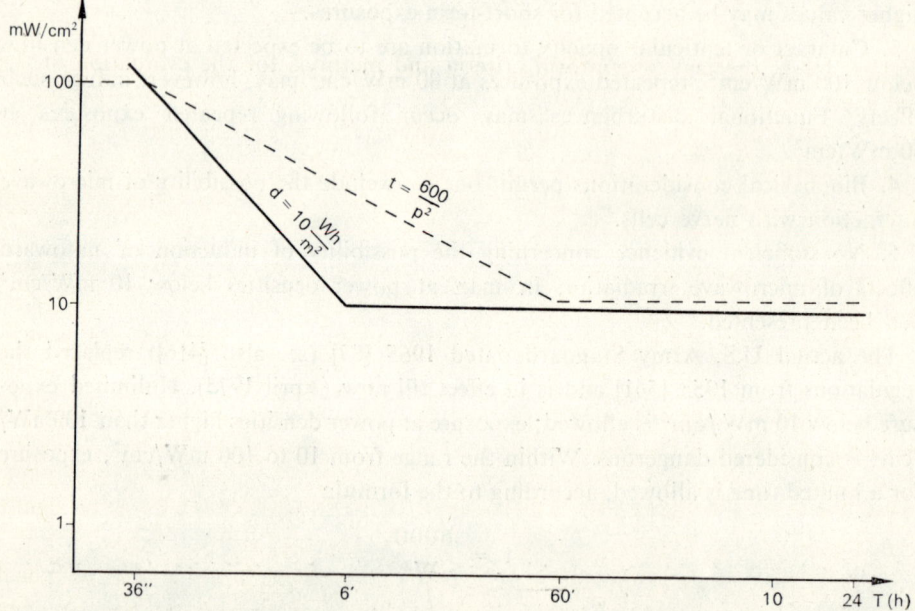

Fig. 67. American safe exposure limits; --- U. S. Army, —— C. 95 ANSI.

averaged over 0.1-h periods. During 1 h 60 mW/cm² would be allowed, according to USASI recommendations for 6 min during 1 h; U.S. Army recommendations allow only a single 2-min exposure at 55 mW/cm² during 1 h. On the other hand USASI recommendations allow a 6-min exposure at 10 mW/cm², while the U.S. Army accepts 32 mW/cm² for this period [416]. The present authors feel that these differences may be neglected for adult exposure in a normal standard environment. It is worth mentioning that U.S. Army regulations demand medical examinations of person working in zones where the power density may be higher than 10 mW/cm². These examinations should be carried out before work in such a zone is allowed (initial examination) and repeated periodically. Besides general physical examination, the lens and the retina must be examined, with the pupil dilatated by appropriate drugs.

The U.S. Navy has regulations dating also from the 1965–1966 period [465; 543] cited according to [491]: unlimited exposure to 10 mW/cm² and intermittent exposure to 300 mJ/cm² for 3 s are allowed. This value corresponds to 100 mW/cm² during 3 s. According to the data available to the present authors, "intermittent exposure" was not defined, and no reference to 0.1-h periods can be made. *Glaser* and *Heimer* [179], discussing microwave-hazard surveys aboard U.S. ships, quote a value of 300 mJ/cm² during 30 s, which corresponds to 10 mW/cm²/30 s. Incidentally it may be pointed out that this paper [179] presents didactically the correct procedure for the determination of the safe and hazardous zones, and should be of interest to any reader active in the field of microwave-hazard prevention.

The recommendation given do not present a full picture of the standpoints taken by various American agencies responsible for health protection and industrial health. *Powell* and *Rose* [434] conducted in 1970 an inquiry in 32 of the 52 states. These authors stress the lack of uniform criteria and methods for the evaluation of the extent of microwave hazards. In three of the states, Bell Telephone Laboratories recommendations were used (*Weiss* and *Mumford* [567]; see also *Mumford* [397, 398]):

1. Power levels in excess of 10 mW/cm² are potentially hazardous and personnel must not be permitted to enter areas where major parts of the body may be exposed to such levels.

2. Power levels between 1 and 10 mW/cm² are to be considered safe only for incidental, occasional, or casual exposure, but are not permissible for extended exposure.

3. Power levels below 1 mW/cm² are safe for indefinitely prolonged exposure.

These recommendations were based initially on the results of animal experimentation (cataractogenic power density of 100 mW/cm², lethal effects at 50 mW/cm²). The value of 1 mW/cm² is based, however, also on considerations of thermal (temperature) effects. These considerations are presented in detail by *Mumford* in a paper reprinted in three publications, which are all cited [398-400] because of the significance of this work. The basic conclusions may be presented in two points:

1. During exposure at 10 mW/cm², the total incident energy being absorbed, heat corresponding to 57.5 W will be generated in a standard man. This should be com-

pared with 73 to 88 W that correspond to basal metabolic rate and 293 W during moderate effort.

2. During exposure at 1 mW/cm², the incident energy corresponds to 5.75 W, i.e., less than 10 percent of the basal metabolic rate. This may be considered as insignificant and neglected from the point of view of heat generation in the organism.

These statements are true for normal conditions in the environment. It should be remembered, however, that heat dissipation depends upon the velocity of air movement, air temperature, and humidity. A correction for these factors becomes necessary. This correction, called the temperature humidity index THI, may be approximated from the following expression:

$$THI = 1{,}44t + 0.1RH3 + 0.6$$

where t = air temperature,
RH = relative humidity.

The approximation is better at higher RH, worse at lower values. At RH equal to 30 percent the error may amount to 3 percent and diminishes to 1 percent at RH of 50 percent the values of the THI being within the range of 70 to 80. One may also determine the THI empirically by making air-temperature measurements using two thermometers, one with a wet and a second with a dry bulb. Then

$$THI = 0.4t_d + t_w + 15 \text{ (temperature °F)}$$

or

$$THI = 0.72t_d + t_w + 40.6 \text{ (temperature °C)}$$

where t_d = temperature reading, dry bulb,
t_w = temperature reading, wet bulb.

Mumford proposed to introduce a correction factor for the THI in the values accepted as safe exposure limits.

To avoid more involved calculations, it may be assumed that with a THI less than 70 the safe exposure limit is 10 mW/cm². Within the range of THI values between 70 and 79, each increase of the THI value by 1 corresponds to a decrease in the safe exposure limit also by 1. This is shown in Figs. 68 and 69, and may be expressed as

$$10 - (THI - 70)$$

when THI values are higher than 79, the safe exposure level remains constant and is equal to 1 mW/cm². This is of course a simplification; details are to be found in the original paper [398-400]. According to *Mumford*, a safety factor of 4 to 5 is obtained by introduction of the correction for the THI. It may be concluded that, from the point of view of the thermal load imposed by microwave irradiation, 1 mW/cm² constitutes a safe exposure limit even in disadvantageous environmental conditions and during physical effort.

Safe Exposure Limits and Prevention of the Health Hazards 175

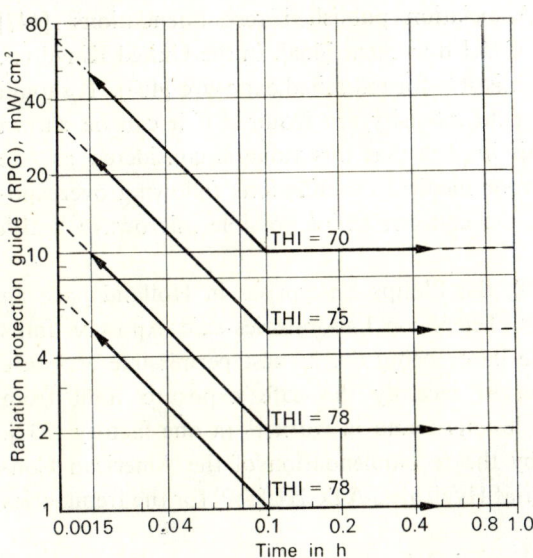

Fig. 68. Exposure time depending on temperature-humidity index. See the text. According to [398].

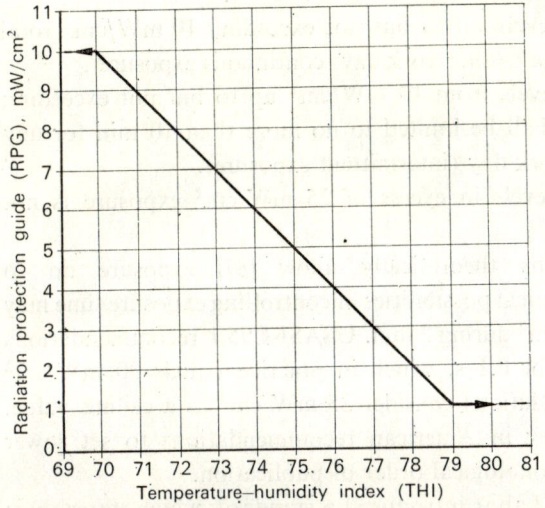

Fig. 69. Dependence of the safe exposure limit on temperature-humidity index. See the text. According to [398].

The preceding considerations are probably the reason for the acceptance by several countries of the safe exposure limit of 1 mW/cm² or 10 mW/cm².

According to *Michaelson* [355] NATO and SHAPE accepted the value of 10 mW/cm². *Swanson et al.* [520] report that French military authorities accepted U.S. Army radiation protection guides recommending, however, a limit of the power density level at public and recreation places to 1 mW/cm². In the German Federal Republic, guidelines [349] that accept the 10 mW/cm² value are considered authoritative. The Canadian Standards Association published recommendations [447] corresponding closely to the USASI-C95.1 document [468]. In the United Kingdom, 10 mW/cm² is also considered as permissible for unlimited exposure [467]. It should be noted that no measurements of field intensity are required if it can be shown that 1 mW/cm² is not exceded. This implies that this value is considered as safe in any conditions. The requirement for medical examinations following overexposure i.e., over 10 mW/cm² stresses the concern about possible microwave health hazards.

According to *Swanson et al.* [520], the Philips Enterprises in Holland have an internal rule [466] accepting the Bell Telephone Laboratories safe exposure limits of 1 mW/cm² and 10 mW/cm², the latter being the highest permissible exposure level. Ericson AB in Sweden lowered recently the safe exposure limit from 10 to 5 mW/cm² in view of retinal lesions found in workers in one factory [337]. These data may be supplemented by the recommendation of the American Conference of Governmental and Industrial Hygienists ACGIH [550] for the frequencies from 100 MHz to 100 GHz:

The threshold limit value for occupational microwave exposure where power densities are known and exposure time controlled is as follows:

1. For average power density levels up to but not exceeding 10 mW/cm² total exposure time shall be limited to the 8-hour work day (continuous exposure).

2. For average power density levels from 10 mW/cm² up to but not exceeding 25 mW/cm² total exposure time shall be limited to no more than 10 min for any 60-min period during an 8-hour work day (intermittent exposure).

3. For average power density levels in excess of 25 mW/cm² exposure is not permissible (ceiling value).

United States Army regulations theoretically allow [87] exposure up to 100 mW/cm², which in view of practical possibilities of controlling exposure time may be considered as equal to 55 mW/cm² during 2 min. USASI-C95.1 recommendations theoretically allow 600 mW/cm² for 0.1 s, which in practice equals 30 mW/cm² for 2 min. The ACGIH recommendations consider 25 mW/cm² as a ceiling value. This points to the steady tendency in American recommendations to set lower ceiling values, according to the chronological order of publication.

In 1975 the U.S. Department of Labor introduced a standard, which states that: ...no employed shall be exposed to power densities exceeding 10 mW/cm²...

All these recommendations are more or less based on human thermal-balance characteristics arising out of *Schwan* and *Piersols* basic papers (biophysical considerations [483] and physiological considerations [484]). These considerations were developed and confirmed in the works of *Schwan* and *Schwan* and *Li* [473, 474, 479-482], *Cook* [91, 92], *Herrick* and *Krusen* [225, 226], *Lehmann et al.* [300-303], *Kalant* [258], *Michaelson* [352-354], and *Seth* and *Michaelson* [490]. A critique on the basis of thermal considerations may be found in *Mumford's* publications [397-400, 567].

Completely different considerations were the basis for the determination of safe exposure limits in the USSR, Poland, and Czechoslovakia.

In the USSR the results of animal experimentation, presented in Chapter 4, and observations on the effects of occupational microwave exposure in man served as a basis for determination of safe exposure limits. According to *Petrov* and *Subbota* [426], the following reasoning was applied to results of animal experimentation: the minimal exposure in terms of functional changes corresponds to 1 mW/cm² during 1 h at 10-cm wavelength; this threshold value extrapolated for a 10-h working day was considered as 0.1 mW/cm². Individual variation in susceptibility, health status, and similar variables was the cause of introducing a tenfold safety margin. The result is 0.01 mW/cm², i.e., 10 μW/cm². Another approach is that of *Gordon* [185], who states simply that the safe exposure limits accepted in the USSR are the results of evaluation of clinical symptoms of microwave professional exposure, as compared to an analysis of working conditions. This was carried out on the material mentioned in Chapter 5, which was analyzed by *Gordon* and her coworkers at the Moscow Institute of Industrial Hygiene and Professional Diseases. According to *Gordon* [182, 186], the tenfold safety coefficient was introduced by setting the safe exposure limits 10 times below the threshold exposure causing any clinical symptoms. On the basis of this reasoning, in 1958 safe exposure limits were defined by law [591], as follows:

1. Exposure levels during the whole working day cannot exceed 0.01 mW/cm² (10 μW/cm²).
2. Exposure for a maximum 2 h/working day must be limited to 0.1 mW/cm² (100 μW/cm²).
3. Exposure for not more than 15 to 20 min/working day at 1 mW/cm² is permissible; protective goggles must be used during such exposure.

It should be added that this concerns professional exposure; the exposure of the population should be limited to 0.001 mW/cm² (1μW/cm²). If the generating equipment operates with a moving beam or antenna circular observation regime, these values may be multiplied by 10 [425]. In the original text it is said that these rules concern exposure to centimeter waves, i.e., 1- to 12-cm wavelengths. Later publications imply that these rules are applied to the whole microwave range, i.e., 300 to 300,000 MHz [425]. Chapter 1 of the Regulations points out that centimeter waves may exert untoward influence on the human organism, expressed by functional disturbances in the nervous and cardiovascular systems and "given sufficiently high exposure levels" also by lenticular opacities. Many protective measures and equipment are used in the USSR [40, 156, 185, 425]. A strictly defined unified method of power density measurements (see *Presman* [438]) was introduced by law. Medical examinations are carried out according to strictly defined rules. Medical contraindications for work in the "microwave-irradiated environment" were determined by the Ministry of Health Protection. These rules are available in English as an appendix to the symposium listed as reference [294]. From the biomedical point of view the law and rules adopted in the USSR are consistent and ensure both unified methods of measurements and medical examinations, as well as unified evaluation of results. The Soviet legislation in this respect may serve as an excellent example of logical, consistent, and comprehensive rules, where all pertinent points are defined and determined. This set of rules served as a model for later Polish and Czechoslovakian ones, which are in consequence inherently logically and consistently construed. In the present authors' opinion, even if the numerical values of safe exposure limits may be discussed, the whole set of USSR rules and legislative measures should serve as a model for similar legislation in other countries. The principal advantages are the following:

1. Safe exposure limits determined separately for professional and incidental (general population) exposure.

2. Unified methods of measurements, measuring equipment, and evaluation of results for hygienic purposes prescribed by law.

3. Unified methods of medical examinations and evaluation of results obtained.

4. Determination of who is responsible where and when for safeguarding the realization of the particular rules and laws.

It may be added incidentally that the Soviet Union was the first country to introduce safe exposure limits for lower frequencies.

The Czechoslovakian safe exposure limits were discussed in English in several publications by *Marha et al.* [336] and *Marha* [332, 333]; see also [520]. For CW exposure 25 μW/cm^2 and for PW exposures 10 μW/cm^2 are accepted as safe. Modernization of these rules introduced a time-weighted average and differentiation between professional and general population (incidental) safe exposure limits. Guide values based on multiples of energy flow per unit area and time were introduced 8-h working day for professional exposure and 24 h for the general population. For professional exposure permissible exposures cannot exceed 200 (guide number) for CW microwaves, i.e., 0.025-mW/cm^2, and for PW guide number 80, i.e., 0.010-mW/cm^2 average density during an 8-h working day. The guide numbers for the general population for the 24-h period are 60 for CW and 24 for PW irradiation, which corresponds to 10-times-lower average exposure limits (0.0025 and 0.001 mW/cm^2). The permissible exposure levels are calculated according to the following formula: guide number (GN) is equal to or more than exposure level (N) in microwatts per square centimeter multiplied by time (t) in hours; i.e.,

$$GN \geq N \cdot t$$

According to this formula, a CW exposure to 1.6 mW/cm^2 or PW exposure to 0.64 mW/cm^2 during 1 h/working day are permissible, which is distinctly higher than the values accepted in the Soviet Union. One wonders what are the values admitted for 1- or 2-min periods. No answer could be found in the available published materials. Details on measuring methods and equipment can be found in the English translation of the monograph by *Marha et al.* [336]. It may be added that continuous generation is defined as operation with the ratio of on-off time greater than 0.1. The maximum peak power is limited for human occupational exposure to 1 kW/cm². The overall impression is that the values accepted in Czechoslovakia may be compared for short periods (2 or 10 min) of exposure to the American USASI or ACGIH recommendation. The value of 1 kW/cm² may pertain, and probably does, to peak pulse power, although the wording is simply "the power density level should not exceed in any moment 1 kW/cm²." According to *Marha* [334] his investigations demonstrate that a difference in the interaction of PW and CW microwaves with the nervous system exists. No details were published. The biomedical considerations on which the Czechoslovakian permissible levels were based could not be found in available literature. In Czechoslovakia safe exposure limits were also determined for lower frequencies in the 30-kHz and 300-MHz range [336].

In Poland the same safe exposure limits as those in the USSR were initially accepted in 1961 [460]. In 1963 additional comments on interpretation were introduced [364]; see also [378, 380]. These comments introduced the notion of effective irradiation time defined by the following expression:

$$t_{ef} = \frac{BW_{ef}}{360°} t_p$$

where t_{ef} = effective irradiation time,
t_p = time of emission of microwaves,
BW_{ef} = effective beam width in degrees,

i.e., beam width at the level where the power density decreases to one half the initial value far-field zone; special formulas are prescribed for the near-field zone (calculated beam width). These interpretations were used for the determination of safe exposure limits to microwaves generated by equipment with a moving beam or antenna (intermittent exposure). In view of the difficulties of solving all the doubts arising out of practical situations, new safe exposure limits were proposed. These were based on detailed discussions of Soviet, Czechoslovakian, and American findings, radiation protection guides, and standards and rules, as well as on an analysis of the health status of personnel professionally exposed to microwaves (statistical epidemiological analysis, as compared to normal incidental exposure [104]). This new proposal was accepted and introduced by laws passed by the Council of Ministers [461] and the Minister of Health and Social Welfare [597]. A distinction between professional and incidental (general population) exposure was made. For the latter the Soviet Union values were accepted: 10 and 100 mW/cm² for continuous (stationary field) and intermittent (nonstationary field) exposure, respectively. These values, corresponding to 0.1 and 1 W/m² in the SI system, were accepted as the upper limits of a safe zone, where habitation or any other human exposure is unrestricted. The general population consists among others of children, juveniles, pregnant women, and individuals with various diseases. The results of animal experimentation and clinical findings (see Chapters 4 and 5) make extreme caution necessary, and in view of uncertainties concerning the mechanism of the interaction of microwaves with living systems, a wide safety margin was introduced. The values determining the upper limit of the safe zone may be said to be based on an "uncertainty coefficient". Experience with healthy adult (professional) exposure [104] determined the upper limits of the remaining zones, i.e., intermediate, hazardous, and dangerous zones.

The actual Polish regulations [461] require that any candidate for work necessitating exposure to microwaves must undergo medical examinations and obtain a medical certificate of fitness. The same requirement is made with respect to candidates for schooling in professions, that will necessitate future exposures to microwaves. Periodic (once a year) medical examinations of microwave workers are compulsory. These examinations include general medical, neurological, ophthalmological, blood, and urine examinations and a chest X-ray. EEG and ECG are deemed desirable on initial examination mainly for future reference. Contraindications for work in a microwave-irradiated environment are determined by an order of the Minister of Health and Social Welfare [596]. This ensures adequate medical health surveillance. Measurement of mean power density in watts per square meter serves for determination of the three zones outside the safe one, i.e., protective zones. A distinction between exposure to stationary and nonstationary fields was made (for definitions see Chapter 2). In conditions of exposure to stationary fields particular zone limits are determined by the following values:

1. Safe zone: the mean power density cannot exceed 0.1 W/m^2; human exposure is unrestricted.

2. Intermediate zone: minimal value of 0.1 W/m^2, upper limit of 2 W/m^2, occupational exposure is allowed during a whole working day (in principle, 8 h may be extended to 10 h for on-duty personnel).

3. Hazardous zone: minimal value of 2 W/m^2, upper limit of 100 W/m^2; the occupational exposure time per 24 h is determined by the formula

$$t = \frac{32}{p^2}$$

where t = permissible exposure time,
p = mean power density in watts per square meter.

4. Dangerous zone: mean power density in excess of 100 W/m^2 (10 mW/cm^2); human exposure is forbidden.

For exposures to nonstationary fields, i.e., intermittent exposure as defined in Chapter 2, the following values were adopted:

1. Safe zone: mean power density does not exceed 1 W/m^2 (0.1 mW/cm^2).

2. Intermediate zone: minimal value of 1 W/m^2, upper limit of 10 W/m^2; occupational exposure is allowed during a whole working day, as in item 2 for stationary fields.

3. Hazardous zone: minimal value of 10 W/m², upper limit of 100 W/m²; the occupational exposure time per 24 h is determined by the formula

$$t = \frac{800}{p^2}$$

where the symbols t and p designate the same values as in the preceding formula.

4. Dangerous zone: mean power density in excess of 100 W/m² (10 mW/cm²); human exposure is forbidden.

The cited law fixes the responsibilities for health surveillance, supervision of working conditions, and the manner of carrying out the measurements (in principle each 3 years or after changes in equipment or its position). The main responsibility for decisions on permissibility of working conditions rests with Sanitary-Epidemiological Stations of the Public Health Service. The necessity for further research and changes in rules is stressed. Newly designed equipment must be evaluated by The Ministry of Health and Social Welfare before its production and/or installation is allowed. The installation of microwave equipment requires a permission from the Sanitary-Epidemiological Station of the province. The Polish set of laws, recommendations, and protection guides is characterized by the same consistency as the Soviet one. The measurements are carried out using similar methods and equipment as in the USSR and Czechoslovakia (see [336, 378, 438, 574]). The English translations of the monographs of *Presman* [438] and *Marha et al.* [336] contain all the pertinent information. Figures 70, 71, and 72 illustrate the Polish safe exposure limits.

In the German Democratic Republic a standard (DDR-Standard TGL 22 314,

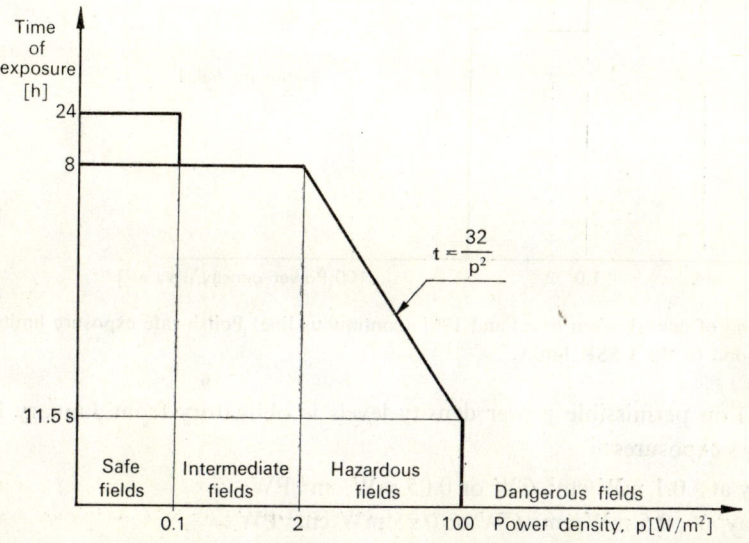

Fig. 70. New Polish safe exposure limit for stationary fields 0.3–300 GHz. See the text.

182 Safe Exposure Limits and Prevention of Health Hazards

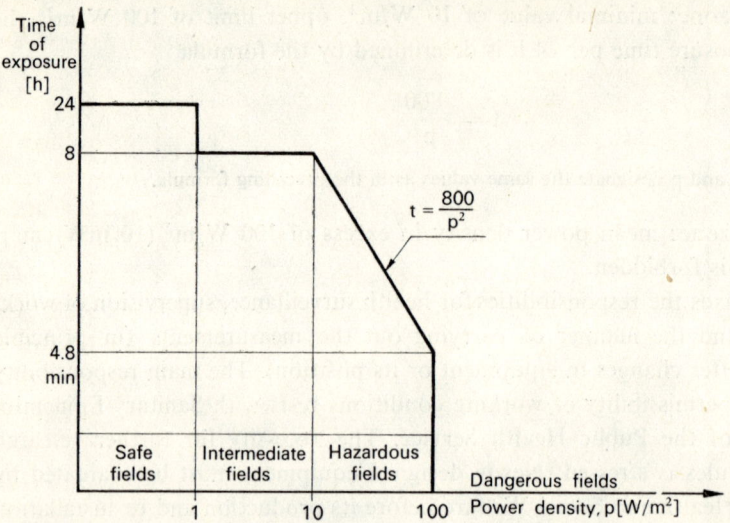

Fig. 71. New Polish safe exposure limit for nonstationary fields 0.3–300 GHz. See the text.

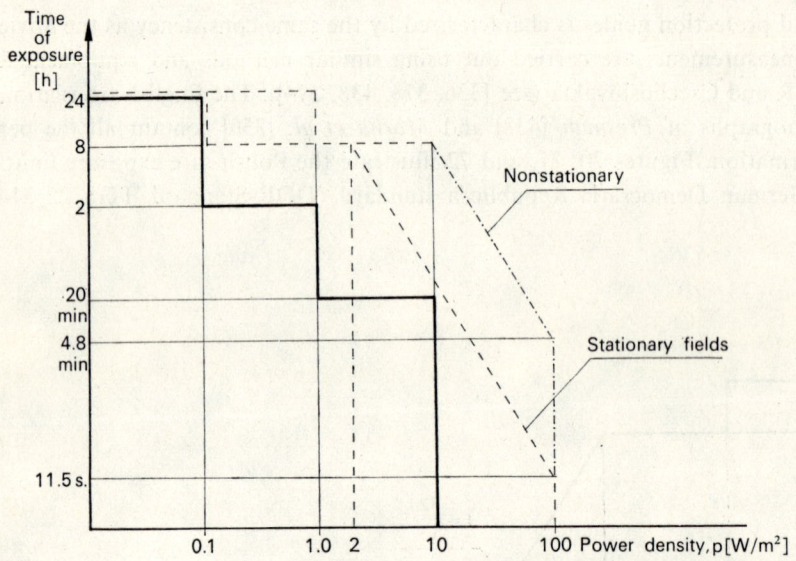

Fig. 72. Comparison of new (broken lines) and 1961 (continuous line) Polish safe exposure limits. The latter correspond to the USSR limits.

Gruppe 963601) on permissible power density levels is obligatory from January 1, 1973. This allows exposures:

up to 8 h/day at   0.1 mW/cm² CW or 0.05 mW/cm² PW
up to 3 h/day at   0.5 mW/cm² CW or 0.25 mW/cm² PW
up to 20 min/day at 1.0 mW/cm² CW or 0.5 mW/cm² PW.

Finally, a few words must be devoted to microwave ovens. These offer obvious economic advantages; moreover, it is claimed at the food cooked in microwave ovens is more easily digested than that prepared in a traditional manner [266, 267]. By definition [422], microwave ovens are used by persons without adequate technical schooling, such as housewives, and vendors at food stalls. It is difficult to imagine a means of adequate technical maintenance and hazard control. At the same time, microwave ovens may be a source of exposure to microwaves among the general population, mainly children and women. It is understandable that this problem has caused much concern and discussion. A voluminous literature on microwave ovens and the prevention of the nonintended radiation hazards involved exists [266, 267, 422, 494, 503, 550, 568]; (see also current numbers of Electronics, Journal of Microwave Power, and Proceedings of the IEEE). From the point of view of this survey, only the permissible level of radiation (leakage) is pertinent. According to the U.S. rules [422], the mean power density level at a distance of 5 cm cannot exceed at any point 1 mW/cm^2 at the moment when a new microwave oven is sold. During use this value may augment to 5 mW/cm^2. According to official statements, this value is consistent with actual knowledge on the biological effects of microwaves and may be considered as safe (Environmental Control Administration, U.S. Public Health Service). Evaluating this statement, and keeping in mind the fact that the value of 5 mW/cm^2 cannot be exceeded at a distance of 5 cm, it may be easily calculated that this is in good accord with the order of magnitude of safe exposure levels for the general population in the USSR and Poland.

Comments on Safe Exposure Limits

The determination of safe exposure limits to any artificial factor introduced into the environment is based on consideration of the relationship between the exposure level and the observed or rather demonstrable bioeffects. Three different basic principles may be used:

1. The principle of "zero" interaction: this level is safe; no effects are demonstrable.

2. The principle of maximal comfort: certain signs are observable but no differences between the functional efficiency of the organism in optimal conditions and on exposure are demonstrable.

3. The principle of the limit of physiological compensation: the exposure causes various disturbances and imposes a stress on the compensatory mechanisms. Nevertheless, no irreversible functional impairment and certainly no irreversible structural changes occur, i.e., exposure does not lead to deviations from the statistical norm.

Another distinct requirement, which must be considered separately, is that the exposure must not induce any genetic effects.

It is difficult to determine to what degree these principles were considered in discussions of microwave safe exposure levels; *Michaelson* [354], *Kucia* [288, 289], and *Czerski* and *Piotrowski* [104] consider such aspects, and a detailed discussion concerning ionizing radiation exposure may be found in *Taylor* [538, 539]. It must be said that the decisions as to what constitutes "maximal comfort" or "limit of physiological compensation" levels are in the present state of biomedical knowledge somewhat arbitrary. It is the present authors' feeling that in the USSR the principle of "zero" interaction was adapted, which is certainly the most cautious and biologically reasonable standpoint in respect to a factor causing so many questions and uncertainties. The same principle was adopted for the general population both in Poland and Czechoslovakia, the main reason being that knowledge of the mechanism of the interaction of microwaves with living systems is insufficient. As concerns occupational exposure, i.e., exposure of healthy adults under medical supervision, a principle of "in between" the "maximal comfort" and "physiological compensation" was aimed at. It may be assumed that Polish, Czechoslovakian, and Bell Telephone (*Mumford*) recommendations are comparable. In Poland it was felt that sufficient empirical data on the effects of human exposure exist to warrant restriction of exposure over 1 mW/cm^2. The definition of stationary and nonstationary fields is such that exposure to near-field zones of high-power equipment, even with a moving beam or antenna, will be considered as stationary field exposure, and thus restricted to lower values. The same can be said for exposure to complex multipath fields in closed rooms, where reflections and interference may cause the occurrence of unpredictable "hot" spots. This cautious standpoint has its basis in the papers of *Bowman* and *Wacker* [49, 50, 563, 564]. These authors point out all the difficulties of quantifying hazardous fields and the inadmissibility of the assumption that the situation of a simple plane wave front illuminating a planar, semi-infinite-slab model may be the basis for determination of safe exposure limits. The present authors' feeling is that this unwarranted assumption has caused not only hygienic misunderstandings, but also questionable interpretation of many experimental data.

It must be stressed that recently many new uncertainties have risen. It must be repeated that power density measurements (see [50, 563, 564]) are unsatisfactory in many situations; measurement of the electric field strength would be probably more meaningful. Determinations of field distributions inside biological objects are necessary. It seems also more reasonable and less costly (in every sense) to accept lower values initially and raise them if possible, than to attempt to control a situation which got out of hand.

Finally it should be pointed out that the safe exposure limits may be considered everywhere as only temporary. The main objections, which concern all the existing recommendations, are as follows:

1. Safe exposure limits pertain to the nonionizing radiation spectrum divided into particular frequency ranges on the basis of technical, and not biological considerations 300,000 to 300 MHz, 300 to 30 MHz, 30 MHz to 30 kHz; an attempt should be made to divide this spectrum into ranges on the basis of biological effec-

tiveness, e.g., 300,000 to 30,000 MHz, 30,000 to 4200 MHz, 4200 to 1500 MHz, and other ranges or in any other manner. The present authors do not feel competent to propose any definite division; this should be a matter for broad discussion.

2. The problem of PW generation and safe exposure limits should be investigated. What are the limits of safe exposure to peak pulse power; what role do pulse width and repetition rate play?

3. Do dangerous or safe irradiation cycle rates exist for nonstationary field exposure (intermittent exposure)?

More similar questions may be raised. The report of a WHO Working Group on Health Effects of Ionizing and Nonionizing Radiation (The Hague Nov. 15—17, 1971) published by the WHO Regional Office for Europe contains almost exclusively doubts and questions on nonionizing radiation health effects. An international effort and a multidisciplinarian approach to all these problems are urgently needed.

An extremely important point is that safe exposure limits (exposure standards) should not be confused with equipment performance standards or emission standards.

Safe exposure limits refer to power density levels and permissible whole-body exposure time at a given power density level. Certain levels are considered hazardous even for very short exposure times, and such levels may be termed ceiling values or limits of danger zones. In other words, such standards (limits or radiation protection guides) concern people.

Emission standards (equipment performance standards) refer to the maximum permissible emission limit of various parts of the equipment. The principal aim of such standards is to eliminate what is called here nonintended radiation (leakage). In other words, such standards refer to the equipment and not to people. It is evident that compliance with emission standards implies safe design of equipment. It is also obvious that emission standards considered from the point of view of the prevention of health hazards are the best solution imaginable. It should be remembered, however, that only nonintended radiation emission may be limited by equipment performance standards. Intended radiation emission depends on the intended use of the equipment, and limitations set below certain requirements of emitted intended radiation power density would render the equipment useless.

It follows that exposure standards refer mainly to intended radiation and may serve to determine restricted zones, where the presence of man should be limited to short periods only or completely forbidden. No emission standard limiting intended radiation is known to the present authors. It may be supposed that in future such standards, or rather international agreements, may be needed, as it may be supposed that a limit to the average electromagnetic pollution of the atmosphere exists.

Emission standards limiting nonintended radiation leakage should ideally tend to eliminate any such radiation. This may be, impracticable, however, from the technical or economic point at the present time. The tendency to eliminate all nonintended radiation is reflected in the U.S. Department of Health, Education, and

Welfare standard concerning microwave ovens. This standard specifies that before first-user purchase the maximum permissible leakage is 1 mW/cm^2 at a distance of 5 cm. With use, this leakage may increase to 5 mW/cm^2 at the same distance. The standard specifies additionally what safety devices should be introduced to ensure compliance with these emission values. The resulting exposure was analyzed by *J. M. Osepchuk, R. A. Foerstner,* and *D. R. McConnell* in a paper, (Computation of Personnel Exposure in Microwave Leakage Fields and Comparison with Personnel Exposure Standards), presented during the International Microwave Power Institute Symposium at Loughborough, Sept. 10–14, 1973. The data presented by these authors indicate that microwave ovens which conform to this standard are a source of exposures within the USSR safe exposure limits.

There remains only the desire that similar emission standards be introduced in respect to any nonintended radiation emission by any part of any microwave equipment.

Prevention of Health Hazards

The analysis and prevention of health hazards has been the subject of many specialized publications, easily accessible in several languages. The reader is referred to the entry health hazards in the subject index appended to the reference list. Only a few of the most important points will be mentioned here. Evidently of greatest importance are appropriate safe exposure limits and their enforcement. It must be stressed that the greatest risk which is often overlooked, is nonintended radiation exposure (leakage). In this respect it is useful to remember the rule that equipment in a state of good repair and used for its intended purpose and according to instructions is safe. Any equipment in a state of diprepair or not so used may be, and usually is, extremely dangerous.

Occupational exposure hazards may be reduced greatly if only healthy adults with recognized health certificates are admitted to work and undergo periodic medical examinations. Any organic or pronounced functional neurological disturbances and/or diseases or head trauma history are contraindications for microwave exposure. In Poland only persons with lens transparency grades 1° and 2° are admitted as candidates for schooling in a profession that will involve microwave exposures; grade 3° in a microwave worker requires observation but does not constitute contraindication, as do grades 4° and 5° [602, 604].

Any neoplastic disease or disorders of the blood-forming system are a contraindication. Blood diseases may be a temporary contraindication till return to normal. The same concerns various disorders of internal organs, such as active stomach or duodenal ulcers or endocrine disorders. These are left for individual decisions of the responsible physician, with special attention being paid to the thyroid.

It should be emphasized that adequate and careful periodic medical examinations and constant medical surveillance are considered as the main safeguard in the prevention of health damage during occupational exposure. Specialistic neurological, ophthalmological, medical, and routine laboratory examinations are obligatory; EEG and ECG are made as indicated.

Protective clothing is used in the United States [179], USSR [425], Poland [99], and Czechoslovakia [336]. No details on design have been published. Protective goggles are also used [99, 145, 336, 425, 427, 438]. It should be remembered that all such equipment may be dangerous if defective (rents, tears, etc.). Being reflecting screens, protective clothes and goggles disturb the configuration of the field and should be considered as "secondary radiation sources" dangerous to the unprotected neighbor. In closed rooms where microwave sources are used, anechoic screens should be installed whenever possible.

More data can be found in the English translations of the monographs by *Marha et al.* [336], *Petrov* [425], and *Presman* [438]. The very useful compilation by *Tell* should be pointed out especially to the American reader [546].

Chapter 7 Final Comments

The authors are fully aware that this survey contains many controversial statements. The reader is once more cautioned to adopt a critical attitude.

Many data are incomplete, simply because no complete descriptions of experiments or, even more so, of human exposure conditions have even been published. It is evident that more controlled experiments and studies are needed. These should be planned so as to make full publication of data possible. Research programs, undertaken without any bias or preconceived ideas, similar to that proposed by *Mills* [372], are a necessity.

The need for multidisciplinarian approaches must be once more emphasized. The problems of microwave bioeffects and health hazards cannot be solved by biologists, physicians, physicists, or electronic engineers working alone. Teamwork is required.

The main obstacles for further advancement of knowledge are the lack of clarity on the primary interaction of microwaves with living matter, insufficient attention to the chain of events leading to secondary early and delayed (late) effects, and the lack of adequate standardized measuring methods and equipment. As concerns the primary interaction, it should be stressed once more that the approaches based on *Debye's* theory and quantum mechanical considerations are legitimate; the latter seem to be more promising at present. Effects on the subcellular and cellular level were till now insufficiently investigated. Special attention should be paid to chromosomal and possible genetic effects. A more complete physiologic approach to secondary effects including those which are secondary, temperature-dependent phenomena, e.g., metabolic effects is needed. It does not suffice to state simply that an effect is thermal and consider the question as solved. Consequences for nervous system function and endocrine system function, as well as for synchronization of the functions of various body organs and systems, must be analyzed. Possible microwave effects on biologic rhythms and their synchronization must be considered. A "vegetative neurosis" or "asthenic" clinical syndrome may well be a consequence of disturbances in regulation and integration of the functions of various body parts.

The final conclusion is that at present more questions can be posed than answers found. It should also be kept in mind that microwaves are only a small segment of the whole nonionizing radiation spectrum. It is the authors' feeling that lower--frequency-radiation bioeffects pose similar questions and perhaps constitute a more serious and ubiquitous health hazard and environmental pollutant than do microwaves.

It is the firm conviction of the present authors that only an international effort based on broad collaboration and unbiased discussions and exchange of opinions may bring a satisfactory solution.

After this book was written three important meetings on microwave bioeffects were held. Chronologically, the first was the Symposium of the International Microwave Power Institute held at Laughborough University in England, September 10–14, 1973. The papers presented during this symposium were published successively in the 1974 issues of the Journal of Microwave Power. The authors would like to draw special attention to the paper by *Leach, Ginns,* and *Rugh,* who demonstrated the teratogenic effects of microwave exposure on pregnant mice (see [605] p. 36). Chromosomal changes in the bone marrow of the offspring of irradiated mice were demonstrated, a finding which is difficult to explain.

In October 1973 an international symposium on "Biologic Effects and Health Hazards of Microwave Radiation" was held in Warsaw under the joint sponsorship of the World Health Organization, the U.S. Department of Health, Education, and Welfare, and the Scientific Council to the Minister of Health and Social Welfare of the Polish Poeple's Republic. This symposium was the first meeting on this subject with large international representation; 69 scientists from 12 countries participated. The proceedings of this symposium was published by Polish Medical Publishers in 1974 [605]. For details, the reader is referred to this book. Only principal points will be mentioned here.

The principal outcome of this symposium seems to be the general agreement that more international collaboration and exchange of information on microwave bioeffects are needed. The inadequacies of translation services, leading to numerous misunderstandings, were pointed out.

The general effects of microwave radiation were discussed, which led to a proposal to divide microwave intensities into three ranges, presented jointly by *Gordon* and *Schwan.* It was also pointed out that an apparent discrepancy exists between the theoretical explanations available, known effects on molecules and isolated components, and the effects observed in complex biosystems. This may arise from insufficient knowledge of the biophysic of such systems, as well as from insufficient knowledge on the primary interaction of microwaves with living matter.

A special session devoted to the effects on the nervous system and behavior demonstrated the importance of research on this subject. Attempts at pharmacological analysis of microwave effects on the nervous system, such as the studies of *Servantie et al.* and *Barański* and *Edelwejn* were considered a promising approach. The importance of electrophysiological studies was stressed. Further progress in this domain depends, however, on the development of electrodes that can be used for recording during microwave exposure without causing artifacts.

Effects on the molecular and cellular level were discussed. Special attention should be paid to the paper by *Rugh* and coworkers, who confirmed the teratogenic effects of exposure of pregnant mice. *Czerski* and coworkers presented experiments indicating that the reaction of the lymphocytic system may be used as an indicator for subtle responses of the organism to low-dose microwave exposure. Particular attention should be paid to effects that depend on modulation during exposure to microwave of the same frequency and mean power density.

The session on the measurement of microwave radiation served only to confirm the inadequacy of present-day measuring equipment and methods. Progress in this field is to be expected in the near future. It should be kept in mind, however, that measurements made in air without the objects to be exposed do not reflect the true field distribution after the objects (animals, men) are introduced into the field. *Beischer* demonstrated that reflections may cause a peculiar standing-wave pattern around the exposed subject. In view of that, if one object is irradiated in given exposure conditions, the introduction of a second object changes radically the field distribution and exposure conditions cease to be comparable. This is particularly important in the evaluation of the exposure of freely moving objects, animals in certain types of experiments, or microwave workers in occupational exposure conditions.

Another difficulty to overcome is the problem of the relationship of the exposure to absorbed energy dose. The biological effectiveness of the same power density exposure doses of microwaves of various frequencies may vary greatly depending both on the differences in absorbed energy dose and its spatial distribution within the body. The next point to resolve is how to evaluate the additive effects of simultaneous exposure to various microwave frequencies and in many instances RF frequencies. This is a point of great practical importance because of the frequent occurrence of such situations in occupational exposure conditions.

The session on occupational exposure effects and the public health aspects of microwave radiation served mainly to point out the necessity of conducting carefully controlled epidemiological studies. The problem of microwave cataracts and lenticular opacities was discussed; the general impression seemed to be that, provided safe exposure limits are kept, no microwave cataracts are to be expected and the incidence of lenticular opacities among microwave workers should not differ from that in comparable age groups.

The symposium conclusions and recommendations were as follows:

"The widespread and increasing use of microwave power has greatly increased the possibility of exposure of both occupational and general population groups in many countries of the world. Protection measures and health and safety standards have varied widely in different parts of the world mainly because of differences in approaches, findings, and interpretations. Insofar as possible, it is imperative to resolve or remove obstacles to a common understanding of the scientific basis for protective measures.

The accomplishments of this memorable international symposium, in terms of meaningful exchange of data, frank discussion of viewpoints, enthusiastic interest in opportunities for collaborative undertakings, and the recommendation that follow, constitute a significant step toward the advancement of knowledge in the field:

1. To promote international coordination of research on the biologic effects of microwave radiation there should be a continuing exchange of information, improved efficiency of translation services, exchange visits, and closer collaboration in research projects and publications.

2. A program concerned with nonionizing radiation should be developed by an international health agency that could exert leadership in this field and facilitate communication among scientists. It was hoped that the World Health Organization would assume this responsibillity.

3. Every effort should be made to establish internationally acceptable nomenclature and definitions of physical quantities and units and to standardize measurement techniques and dosimetry. It is recommended that an international group be established to work out procedures for achieving these objectives.

4. To achieve a more uniform approach to the discussion of mechanisms underlying biologic effects, it is proposed that microwave intensities be divided into three approximate ranges as follows:

  a. The range above 10 mW/cm^2 in which distinct thermal effects predominate.
  b. The range below 1 mW/cm^2 in which thermal effects are improbable.
  c. An intermediate range in which weak but noticeable thermal effects occur as well as direct field effects and other effects of a microscopic or macroscopic nature, the details of which have not yet been clarified.

The limits of these ranges have not yet been determined. They may differ for various species and may also depend on a variety of parameters, such as modulation and frequency.

5. In view of the importance of electrophysiological recording for studies of microwave effects there is need for the development of new electrode systems and integrated electrode signal amplifying systems capable of use and full operation during microwave exposure.

6. Further biologic, medical, epidemiologic and biophysical studies are needed to improve understanding of the interactions of microwave radiation with biologic systems and clarify the risks that may be associated with microwave exposure. Specific attention should be given to the following:

  a) investigation into the occurrence of cumulative effects and delayed effects;
  b) study of low intensity effects;
  c) determination of possible threshold values;
  d) study of combined effects of radiation and other environmental factors;
  e) investigation of differential radiation sensitivity as a function of organ system and age or intrauterine development;
  f) study of effects occurring at the molecular level;

g) study of effects related to cellular transformations;

h) determination of absorbed energy dose and its spatial distribution.

The desirability of conducting similar investigations in the radiofrequency range was emphasized".

In connection with this symposium it should be mentioned that the Regional Office for Europe of the World Health Organization started a long-term program concerning nonionizing radiation protection. It may be expected that several WHO reports concerning definitions, methods, biological effects, and safe exposure limits as related to health hazards from microwave radiation will be published. These publications should be looked for.

The New York Academy of Sciences Conference on Biological Effects of Nonionizing Radiation, New York, Feb. 12–15, 1974, [606] comprised six sessions:

1. Electromagnetic radiation effect on the nervous system.
2. Electromagnetic radiation effect on special senses.
3. Biochemical and biophysical effects.
4. Electromagnetic radiation effects on genetics and development.
5. Behavioral effects of electromagnetic radiation.
6. Dosimetry of electromagnetic radiation.

The first and second sessions concerned the effects of modulated fields on the nervous system, with particular attention to morphological and electrophysiological studies, and effects on the auditory system and the eye. The third session comprised papers devoted to electromagnetic radiation effects on the cellular and subcellular level. It should be stressed that this conference was the first one during which sufficient material was found to organize a special session on genetic effects and influence on development. Nevertheless, this important field needs still further studies, present knowledge on this subject being still in the initial stage. The fifth session comprised studies complementary to those presented during the first one. The last session concerned mostly the problems of the distribution of absorbed energy within biological objects. All papers presented during this conference appeared in a special issue of the Annals of the New York Academy of Sciences [606].

In addition, the present authors would like to draw the reader's attention to a monograph on nonionizing radiation bioeffects by *S. M. Michaelson,* which will be published during 1977 in the United States. Because of certain differences in approach, this book should allow the reader to evaluate the present monograph in a more objective manner.

References

1. *Abrikosov*, I. A.: Impulsnoje elektritcheskoje polje ultravysokoj tchastoty. Medgiz. Moscow, 1958. (Pulsed electric UHF fields)
2. *Addington, C., Fischer, F., Neubauer, R., Osborne, C., Sarkees, Y., Swartz, G.*: Studies on the biological effects of 200 megacycles. In [418], p. 189.
3. *Addington, C. H., Osborne, C., Swartz, G., Fischer, F., Sarkees, Y. T.*: Thermal effects of 200 megacycles (CW) irradiation as related to shape, location and orientation in the field. In [518], p. 10.
4. *Adler, R. B., Chul, J., Fano, R. M.*: Electromagnetic energy transmission and radiation. John Wiley & Sons, Inc. New York, 1960.
5. *Anne, A., Saito, M., Salati, O. M., Schwan, H. P.*: Penetration and thermal dissipation of microwaves in tissues. Univ. of Pennsylvania. Philadelphia. ASTIA Doc. **284**, 981, 1962.
6. *Anne, A.*: Scattering and absorption of microwaves by dissipative dielectric objects: the biological significance and hazard to mankind. Ph. D. dissertation. Univ. of Pennsylvania. Philadelphia. ASTIA Doc. **408**, 997, 1963.
7. *Antczak, K., Minecki, L., Więcek, E.*: Zmiany aktywności cholinesterazy we krwi i narządach wewnętrznych myszy poddawanych działaniu mikrofal (pasmo S). Abst. Ref. VI Krajowa Konferencja Medycyny Pracy. Gdańsk, June 14–17, 1962. p. 10.
[Cholinesterase activity changes in the blood and internal organs of white mice subjected to microwaves (S band)]
8. *Auerswald, W.*: Temperaturtopographische Untersuchungen zur Frage der Wirkung von Kurzwellendurchflutung des Zwischenhirns. Wien. Zt. Nervenheilkunde **4**, 273, 1952.
(Investigations on temperature topography bearing on the question of influence of shortwave irradiation of the mesencephalon)
9. *Austin, G. M., Horwath, S. M.*: Production of convulsions in rats by exposure to ultrahigh frequency electrical currents (radar). Am. J. Med. Sci. **218**, 115, 1949.
10. *Austin, G. M., Horwath, S. M.*: Production of convulsions in rats by high frequency electrical currents. Am. J. Phys. Med. **33**, 141, 1954.
11. *Bach, S. A.*: Biological sensitivity to radiofrequency and microwave energy. Federation Proc. **24**, 3, 1965.
12. *Bach, S. A., Baldwin, M., Lewis, S.*: Some effects of ultrahigh frequency energy on primate cerebral activity. In [518], p. 82.
13. *Bach, S. A., Luzzio, A. J., Brownell, A. S.*: Effects of R. F. energy on human gamma globulins. In [80], p. 117.
14. *Baillie, H. D., Heaton, A. G., Pal, D. K.*: Effects of metallic inductions in a microwave field. Non-ionizing Radiation **1**, 120, 1969.
15. *Baillie, H. D.*: Thermal and non-thermal cataractogenesis by microwaves. Non-ionizing Radiation **1**, 159, 1970; see also [82], p. 59.
16. *Baillie, H. D., Heaton, A. G., Pal, D. K.*: The dissipation of microwaves as heat in the eye. Non-ionizing Radiation **1**, 164, 1970; see also [82], p. 85.
17. *Balleste, G. F., Zingano, A. G.*: Medical aspects of the operation of radar and similar apparatus. Riv. Med. Aeron. **14**, 3, 1962.
18. *Barański, S.*: Badania nad wbudowywaniem się związków znakowanych izotopami promieniotwórczymi w ośrodkowym układzie nerwowym w przebiegu niedotlenienia wysokościowego. Med. Lotnicza **8**, 16, 1962.

(Investigations on radioisotope tagged substances incorporation in the central nervous system in the course of high-altitude hypoxia)
19. *Barański, S.*: Histological and histochemical effects of microwave irradiation on the central nervous system of rabbits and guinea pigs. Am. J. Phys. Med. **51**, 182, 1972.
20. *Barański, S.*: Badania biologicznych efektów swoistego oddziaływania mikrofal. Inspektorat Lotnictwa. Warsaw, 1967.
(Investigations on specific microwave bioeffects)
21. *Barański, S.*: Biologiczne działanie mikrofal. In [574], p. 78.
(Biological effects of microwaves)
22. *Barański, S.*: Wpływ mikrofal na odczyn układu białokrwinkowego. Acta Physiol. Polon. 22, 898, 1971; see also Aerospace Med. **42**, 1196, 1971.
(Influence of microwaves on white blood cell response)
23. *Barański, S., Czekaliński, L., Czerski, P., Haduch, S.*: Recherches experimentales sur l'effet mortel de l'irradiation des ondes micrometriques. Rev. Med. Aeron. **2**, 108, 1963.
(Experimental investigations on lethal effects of microwave irradiation)
24. *Barański, S., Czerski, P.*: Badania zachowania się składników upostaciowanych krwi u osób zatrudnionych w zasięgu mikrofal. Lek. Wojsk. **4**, 903, 1966.
(Investigations on the morphotic elements of blood in persons professionally exposed to microwaves)
25. *Barański, S., Czerski, P., Czekaliński, L., Haduch, S.*: Badania nad wpływem dużej gęstości mocy mikrofal w paśmie 3 cm na ustrój myszy. Med. Pracy **14**, 129, 1963.
(Investigations on the influence of high-power-density microwaves of the 3-cm band on mice)
26. *Barański, S., Czerski, P.*: Swoiste efekty biologicznego oddziaływania mikrofal. Med. Lotnicza **39**, 1972, p. 11.
(Specific effects of biologic interaction of microwaves)
27. *Barański, S., Czerski, P., Janicka, U., Szmigielski, S.*: Wpływ naświetlania mikrofalami na rytm dobowy podziału komórek szpiku. Med. Lotnicza **39**, 1972, p. 67.
(Influence of microwave irradiation on circadian rhythm of bone marrow cell division)
28. *Barański, S., Czerski, P., Szmigielski, S.*: Wpływ mikrofal na mitozę *in vitro* i *in vivo*. Postępy Fiz. Medycznej **6**, 93, 1971.
(The influence of microwaves on mitosis *in vitro* and *in vivo*)
29. *Barański, S., Edelwejn, Z., Kaleta, Z.*: Badania czynnościowe i morfologiczne mięśni poddanych działaniu mikrofal. Med. Lotnicza **24**, 103, 1967.
(Functional and morphologic examinations on microwave irradiated muscles)
30. *Barański, S., Edelwejn, Z.*: Badania skojarzonego działania mikrofal i niektórych leków na czynność bioelektryczną ośrodkowego układu nerwowego u królików. Acta Physiol. Polon. **19**, 37, 1968.
(Studies on the combined effect of microwaves and some drugs on bioelectric activity of the rabbit central nervous system)
31. *Barański, S., Edelwejn, Z.*: Badania elektroencefalograficzne i morfologiczne nad wpływem mikrofal na ośrodkowy układ nerwowy. Acta Physiol. Polon. **18**, 517, 1967.
(Electroencephalographic and morphological investigations on the influence of microwaves on the central nervous system)
32. *Barański, S., Ostrowski, K., Stodolnik-Barańska, W.*: Experimental investigations on the influence of microwaves on thyroid function. Acta Physiol. Polon., **23**, 1973, 608.
33. *Barnothy, M.* (ed.): Biological effects of magnetic fields. Plenum Publishing Corporation, New York, 1964.
34. *Barron, Ch., Love, A., Baraff, A. Y.*: Physical evaluation of personnel exposed to microwave emanations. (See also *Barron, Ch., Barraff, A.*: J. Am. Med. Assoc. **169**, 1194, 1958).

35. Basic radiation protection criteria. National Council on Radiation Protection and Measurements (NCRP) Rep. 39. Washington, D. C., 1971.
36. *Baus, R., Fleming, J. D.*: Biological effects of microwave with limited body heating. In [518], p. 291.
37. *Bavro, G. V., Cholodov, J. A.*: Charakter bioelektritcheskich reakcji kory golovnovo mozga krolikov pri vozdiejstvi SVTCh polia. In Voprosy biologitcheskogo diejstvia SVTCh elektromagnitnogo polia.
    Abstr. Tez. Naucz. Konferen. Leningrad, 1962.
    (Character of bioelectric responses of the cortex to UHF fields in rabbits)
38. *Beischer, D.*: Biomagnetics. Ann. N. Y. Acad. Sci. **134**, 454, 1965.
39. *Beisang, A. A., Mayo, C. H., Pace, M., Lillehei, R. C., Graham, E. F.*: Rapid thawing of extended semen with microwaves. Symposium IMPI. Univ. Alberta Edmonton May 21–23, 1969. Session DB. Abstracts, p. 89.
40. *Belicki, B. M., Knoppe, K. G.*: Zashtchita od izlutchenii pri rabote s generatorami SVTCh. In [294], p. 110. (Radiation protection when working with UHF generators)
41. *Berest, N., Perdriel, J., Colin, J.*: Etude statistique de la fatique visuelle des lecteurs de scope radar. Med. Aeron. **13**, 135, 1958.
    (Statistical study of ocular fatigue in radar operators)
42. *Bereznickaja, A. N., Kazbekov, I. M., Rysina, T. Z.*: Vlijanije mikrovoln netermogennych intensivnosti na generativnuju funkciju myšei i ich potomstvo. Gigiena Truda i Biologitcheskoe Deistvie Elektromagnitnych Voln Radio Tchastot. Symp. Abstr. Moscow, 1972, p. 35.
    (The influence of microwaves at nonthermogenic intensities on generative function in mice and on their offspring)
43. *Bessonova, A. F.*: O vlijanii vnietermitcheskich vozdeistv SVTCh-polja na funkcjonalnoje sostojanie izolirovanego nerva. In Voprosy biologitcheskogo deistvija sverchvysokotchastotnogo (SVTCh) elektromagnitnogo polja. Tezisy Nautchnoi Konferencii. Leningrad, 1962.
    (On the nonthermal effects of UHF fields on the functional state of isolated nerves)
44. *Bielicki, Z., Barański, S., Czerski, P., Haduch, S.*: Analyse des troubles de l'activite professionnelle chez le personnel expose a l'irradiation des ondes micrometriques. Rev. Med. Aeron. **26**, 106, 1963.
    (Disturbances of professional activity in personnel exposed to microwave irradiations)
45. *Belova, S. F.*: Deistvie santimetrovych voln na glaza. Bjul. Eksper. Biol. Med. **41**, 43, 1956.
    (The influence of centimeter waves on eyes)
46. *Birenbaum, L., Grosof, G. M., Rosenthal, S. W., Zaret M. M.*: Effect of microwaves on the eye. IEEE Trans. B.M.E. **16**, 7, 1969.
47. *Birenbaum, L., Kaplan, T. T., Metlay, W., Rosenthal, S. W., Schmidt, H., Zaret, M. M.*: Effect of microwaves on rabbit eye. J. Microwave Power **4**, 232, 1969.
48. *Boni, A., Lotmar, R.*: Temperaturmessungen an thierischen und menschlichen Geweben nach Bestrahlung mit Mikrowellen. Arch. Physik. Therap. **3**, 174, 1952.
    (Temperature measurements on animal and human tissues after microwave irradiation)
49. *Bowman, R. R.*: Quantifying hazardous electromagnetic fields. National Bureau of Standards. Tech. Note, 389. April 1970; see also [82], p. 204.
50. *Bowman, R. R.*: Quantifying hazardous microwave fields: practical considerations. In [82], p. 204.
51. *Boysen, J. E.*: USAF experience with microwave exposure. J. Occupational Med. **4**, 192, 1962.
52. *Boysen, J. E.*: Hyperthermic and pathologic effects of electromagnetic radiation (350 Mc). AMA Arch. Ind. Hyg. Occupational Med. **7**, 516, 1953.
53. *Brody, S. J.*: The operational hazard of microwave radiation. J. Aviat. Med. **24**, 328, 1952.
54. *Brody, S.*: Bioenergetics and growth. Van Nostrand Reinhold Company, New York, 1945.

55. *Busco, R. Comignani, L.*: Nozioni attuali circa gli effetti delle onde radar sugli organismi viventi ed i relativ mezzi di protezione. Parte prima: Principi generali fisici ed effetti fisiopathologici. Riv. Med. Aeron. Spaziale **30**, 469, 1967.
(Observations on effects of radar waves on living organisms and appropriate protective measures. Part I: General physical principles and physiologic effects)
56. *Busco, R., Comignani, L.*: Nozioni attuali circa gli effetti delle onde radar sugli organismi viventi ed i relativi mezzi di protezione. Parte seconda. Riv. Med. Aeron. Spaziale **30**, 718, 1967.
(Observations on effects of radar waves on living organisms and appropriate protective measures)
57. *Bytchkov, M. S.*: O mechanizmie diejstvi SVTCh elektromagnitnogo polia. Abstr. Tez. Nautch. Konf. Leningrad, 1962.
(On the mechanism of SHF electromagnetic field effects)
58. *Bytchkov, M. S.*: Elektroencefalografitchsskoje issledovanija vlijanii SVTCh polia na tchełovieka. Abstr. Tez. Nauczn. Konfer. Leningard, 1962.
(Electroencephalographic studies on the influence of SHF field on man)
59. *Bytchkov, M. S., Syngajevskaja, V. A.*: Eksperymentalnyje dannyje o spravitelnoj effektivnosti oblutchenii v reżimie krugovogo obzora i s fiksirovannoj antenoj. Abstr. Tez. Nautch. Konf. Leningrad, 1962.
(Experimental data on relative effects of irradiation with a moving or stationary antenna)
60. *Bytchkov, M. S., Syngajevskaja, V. A.*: Materiały o vnieteplovom vozdiejstvi polia na cholinergitcheskije sistemy organizma. Abstr. Tez. Nautch. Konf. Leningrad, 1962.
(Materials on nonthermal effects of SHF fields on cholinergic systems of the organism)
61. *Carapancea, M., Popescu, M., Stefan, M., Bengulescu, D., Musetescu, S.*: Tulburarile vizuale si consecintele lor asupra organismului, in conditiile perceptiei la dispozitivul radar. Studii si cercetari de fiziologie (Bucuresti) **6**, 695, 1961.
(Visual troubles and their consequences for the organism from conditions due to radar screen observation)
62. *Carney, S. A., Lawrence, J. C.*: Effect of microwaves at X-band on guinea pig skin in tissue culture. 2. Effect of the radiation on skin biochemistry. Brit. J. Ind. Med. **25**, 229, 1968.
63. *Carney, S. A., Lawrence, J. C., Richetts, C. R.*: Effects of microwaves at X-band on guinea pig skin in tissue culture. 3. Effect of pulsed microwaves on skin respiration biochemistry. Brit. J. Ind. Med. **27**, 72, 1970.
64. *Carpenter, R. L.*: Experimental microwave cataract, a review. In [82], p. 76.
65. *Carpenter, R. L.*: Experimental radiation cataracts induced by microwave radiation. Proc. Second Tri-Serv. Conf. Biol. Effects Microwave Energy. RADC-TR-54, 146, pp., 1958.
66. *Carpenter, R. L.*: Studies on the effects of 2450 megacycle radiation on the eye of the rabbit. Proc. Third Tri-Serv. Conf. Biol. Effects Microwave Energy. RADC-TR-59-140, 279, pp., 1959.
67. *Carpenter, R. L.*: An experimental study of the biological effects of microwave radiation in relation to the eye. RADC-TDR-62-131, p. 51, 1962.
68. *Carpenter, R. L., Biddle, D. K., Van Ummersen, C. A.*: Opacities in the lens of the eye experimentally induced by exposure to microwave radiation. IRE Trans. Med. Electron. **7**, 152, 1960.
69. *Carpenter, R. L., Livstone, E. M.*: Evidence for nonthermal effects of microwave radiation: abnormal development of irradiated insect pupae. IEEE Trans. Microwave Theory and Techniques, vol. MTT-19, 173, 1971.
70. *Carpenter, R. L., Van Ummersen, C. A.*: The action of microwave radiation in the eye. J. Microwave Power **3**, 3, 1968.
71. *Cholodov, J. A.*: O izmienienji elektritcheskoj funkcii kory gołovnogo mozga krolikov pri vozdiejstvi elektromagnitnogo polia SVTCh. Bjul. Eksper. Med. **9**, 42, 1963.

(Electrical functional changes in brain cortex of rabbits under the influence of electromagnetics SHF field)
72. *Cholodov, J. A.*: Vlijanii electromagnitnych i magnitnych poljej na centralnuju nervnuju sistemu. Medgiz. Moskva, 1966.
(The influence of electromagnetic and magnetic fields on the central nervous system)
73. *Cholodov, J. A., Zenina, I. N.*: Vlijanie kofeina na EEG-reakciju na vozdeistvii impulsnogo SVTCh polja na intaktnyi i izolirovannyi mozg krolika. In: O biologitcheskom deistvii elektromagnitnych polei radiotchastot. Moskva, 1964, p. 33.
(On the influence of caffeine on the EEG reaction to SHF fields in intact and isolated rabbit brain)
74. *Cieciura, L.*: Badania histochemiczne nad aktywnością niektórych enzymów w prawidłowej i uszkodzonej gonadzie męskiej szczura białego (thesis). Wojskowa Akademia Medyczna. Łódź, 1964.
(Histochemical investigations on activity of certain enzymes in the normal and damaged male gonad of white rats)
75. *Cieciura, L., Hibner, H., Minecki, L., Andryszak, G.*: Wpływ promieniowania mikrofalowego na mitozę stożka wzrostowego korzonka bobu (*Vicia faba*). Med. Pracy **22**, 211, 1971.
(The influence of microwave radiation on the mitosis in root tips of the bean *Vicia faba*)
76. *Cieciura, L., Karasek, M., Pawlikowski, M., Minecki, L.*: Wpływ mikrofal na ultrastrukturę szyszynki u szczurów białych. Folia Morphol. **28**, 343, 1969.
(The influence of microwaves on the ultrastructure of the pineal gland in white rats)
77. *Cieciura, L., Minecki, L.*: Zmiany histopatologiczne w jądrach szczurów poddanych jednorazowemu i wielokrotnemu działaniu mikrofal (pasmo S). Lek. Wojsk. **38**, 519, 1962.
[Histopathologic changes in testicles of rats subjected to single or repeated microwave irradiation (S band)]
78. *Cieciura, L., Minecki, L.*: Zmiany histopatologiczne w jądrach szczurów poddanych działaniu promieniowania mikrofalowego w stanie hipotermii. Med. Pracy **27**, 507, 1966.
(Histopathologic changes in testicles of hypothermic microwave-irradiated rats)
79. *Cieciura, L., Minecki, L.*: Rozmieszczenie i aktywność niektórych enzymów hydrolitycznych w jądrach szczurów poddanych działaniu mikrofal (pasmo S). Med. Pracy **15**, 159, 1964.
[Distribution and activity of certain hydrolitic enzymes in testicles of rats subjected to microwaves (S band)]
80. *Cleary, S. F.*: Biological effects of microwave and radiofrequency radiation. CRC Critical Rev. Environ. Control, June 1970, p. 257.
81. *Cleary, S. F., Pasternak, B. S.*: Cataract incidence in radar workers. Arch. Environ. Health **11**, 179, 1965.
82. *Cleary, S. F.* (ed): Biological effects and health implications of microwave radiation. Symposium proceedings (Richmond, V., Sept. 17, 1969). U. S. Department of Health, Education, and Welfare. Report BRH/DBE 70-2 (PB 193 858). Rockville, Md., 1970.
83. *Coccorra, G., Blasio, A., Nunciata, B.*: Rilievi sulle embriopatie da onde corte. La pediatria— riv. igiene med. chir. infantia **68**, 7, 1960.
(Remarks on embryopathies induced by short waves)
84. *Cogan, D. G., Donaldson, D. D.*: Experimental radiation cataracts: I. Cataracts in the rabbit following single X-ray Exposure. Arch. Ophthalmol. **45**, 508, 1951.
85. *Cogan, D. G., Fricker, S. J., Lubin, M., Donaldson, D. D., Hardy, H.*: Cataracts and ultrahigh frequency radiation. AMA Arch. of Ind. Health **18**, 299, 1958.
86. *Cogan, D. G., Goft, J. L., Graves, E.*: Experimental radiation cataracts: II. Cataracts in the rabbit following single exposure to fast neutrons. Arch. Ophthalmol. **47**, 584, 1952.
87. Control of hazards to health from microwave radiation. U. S. Department of the Army and the Air Force Rep. TB MED 270/AFM, 161, 1965.

88. *Cook, H.*: A comparison of the dielectric behaviour of pure water and human blood at microwave frequencies. Brit. J. Appl. Phys. **3**, 249, 1952.
89. *Cook, H.*: Dielectric behaviour of human blood at microwave frequencies. Nature **168**, 247, 1951.
90. *Cook, H.*: The dielectric behaviour of some types of human tissues at microwave frequencies. Brit. J. Appl. Phys. **2**, 295, 1951.
91. *Cook, H.*: A physical investigation of heat production in human tissues when exposed to microwaves. Brit. J. Appl. Phys. **3**, 1, 1952.
92. *Cook, H.*: The pain threshold for microwave and infra-red radiation. J. Physiol. **118**, 1, 1952.
93. *Cooper, T., Jellinek, M., Pinakatt, T., Richardson, A. W.*: The effects of pyridoxine and pyridoxal on the circulatory responses of rats to microwave irradiation. Experientia **21**, 28, 1965.
94. *Cooper, T., Pinakatt, T., Jellinek, M., Richardson, A.*: Effects of adrenolectomy, vagotomy, and ganglionic blockade on the circulatory response to microwave hyperthemia. Aerospace Med. **33**, 794, 1962.
95. *Cooper, T., Pinakatt, T., Richardson, A. W.*: Effect of microwave induced hyperthermia on the cardiac output of the rat. Physiologist **4**, 21, 1961.
96. *Cooper, T., Pinakatt, T., Jellinek, M., Richardson, A. W.*: Effects of reserpine on the circulation of the rat after microwave irradiation. Am. J. Physiol. **202**, 1171, 1962.
97. *Corson, D., Lorrain, P.*: Introduction to electromagnetic fields and waves. W. H. Freeman and Company, San Fransisco, 1962.
98. *Crapuchette, P. W.*: Instrumentation for microwave leakage. Non-ionizing Radiation **2**, 15, 1971; see also [82], p. 210.
99. *Czerski, P.*: Voprosy gigieny i ochrony truda pri obsluzivanii generatorov mikrovoln. Med. Lotnicza, 1965. Nr. specjalny, Konferencja Lekarzy Lotniczych Europejskich Krajów Socjalistycznych.
(Hygienic problems and labor protection during work with microwave generators. Special number, Conference of Flight Surgeons of European Socialist Countries, 1965)
100. *Czerski, P.*: Lymphoblastoid transformation induced *in vitro* by microwave irradiation. Preliminary report. IVth Immunology Symposium. Poznań. May 21–22, 1972.
101. *Czerski, P.*: Issledovanie vlijanija mikrovoln na krovetvormiju sistemu. In: Gigiena Truda i Biologitcheskoe Deistvie Elektromagnitnych Voln Radiotchastot. Symp. Abstr. Moscow, 1972, p. 56.
(Investigations on the influence of microwaves on the blood-forming system)
102. *Czerski, P., Barański, S., Siekierzyński, M.*: Microwave irradiation and bone marrow function. III. International Conference on Medical Physics. Göteborg, 1972. Abstr. 39.9.
103. *Czerski, P., Hornowski, J., Szewczykowski, J.*: Przypadek choroby mikrofalowej. Med. Pracy **15**, 251, 1964.
(A case of microwave syndrome)
104. *Czerski, P., Piotrowski, M.*: Założenia do ustalenia dopuszczalnych dawek mikrofal dla ludzi. Med. Lotnicza 39, 1972, p. 127.
(Premises for setting up safe limits for human microwave exposure)
105. *Czerski, P., Witkowicz, J.*: Ochrona przed działaniem mikrofal. In [574], p. 125.
(Protection against microwave hazards)
106. *Daily, L.*: A clinical study of the results of exposure of laboratory personnel to radar and high frequency radio. U. S. Navy Med. Bull. **41**, 1052, 1943.
107. *Daily, L., Wakim, K. G., Herrick, J. F., Parkhill, E. M., Benedict, W. L.*: The effects of microwave diathermy on the eye of the rabbit. A. J. Ophtalmol. **35**, 1001, 1952.
108. *Daily, L., Wakim, K. G., Herrick, J. F. and Parkhill, E. M.*: The effects of microwave diathermy on the eye: an experimental study. J. Ophtalmol. **33**, 1245, 1950.

109. *Daily, L., Wakim, K. G., Herrick, J. F., Parkhill, E. M.*: Effects of microwave diathermy on the eye. Am. J. Physiol. **155**, 432, 1948.
110. *Daily, L., Wakim, K. G., Herrick, J. F., Parkhill, E. M.*: The effects of microwave diathermy on the eye: an experimental study. Am. J. Ophtalmol. **34**, 1137, 1951.
111. *Daily, L., Zeller, K. G., Herrick, J. F., Benedict, W. L.*: Influence of microwaves on certain enzyme systems in the lens of the eye. Am. J. Ophtalmol. **34**, 1301, 1951.
112. *Davis, T. R. A., Meyer, J.*: Use of high frequency electromagnetic waves in study of thermogenetics. A. J. Phys. Med. **178**, 283, 1954.
113. *Deichmann, W. B.*: Results of studies of microwave radiation. In [518], p. 72.
114. *Deichmann, W. B.*: Biological effects of microwave radiation of 24,000 megacycles. Arch. Toxika **22**, 24, 1966.
115. *Deichmann, W. B.*: Annual report of microwave radiation research. RADC Tech. Report **59**, 228, 1959.
116. *Deichmann, W. B.*: Introducing the "irradiation cycle rate" in microwave radiation exposures. Biochem. Pharmacol. **8**, 157, 1961.
117. *Deichmann, W. B., Bernal, E., Stephens, F., Landeen, K.*: Effects on dogs of chronic exposure to microwave radiation. J. Occup. Med. **5**, 418, 1963.
118. *Deichmann, W. B., Bernal, E., Keplinger, M.*: Effects of environmental temperature and air volume exchange on survival of rats exposed to microwave radiation of 24,000 megacycles. Ind. Med. Surg. **28**, 535, 1959.
119. *Deichmann, W. B., Miale, J., Landeen, K.*: Effect of microwave radiation on the hemopoietic system of the rat. Toxicol. Appl. Pharmacol. **6**, 71, 1964.
120. *Deichmann, W. B., Keplinger, M., Bernal, E.*: Relation of interrupted pulsed microwaves to biological hazards. Ind. Med. Surg. **25**, 212, 1959.
121. *Deichmann, W. B., Stephens, F. H., Keplinger, M., Lampe, K. F.*: Acute effects of microwave radiation on experimental animals (24,000 megacycles). J. Occupational Med. **1**, 369, 1959.
122. *Deichmann, W. B., Stephens, F. H.*: Microwave radiation of 10 mW/cm^2 and factors that influence biological effects at various power densities. Ind. Med. Surg. **30**, 221, 1961.
123. *Dęga, K., Klajman, S., Doboszyński, T.*: Wpływ 10 cm fal radaru na ciepłotę ciała i przemianę gazową u szczurów. Biul. Wojskowej Akad. Med. **8**, 557, 1964.
(The influence of 10-cm radar waves on body temperature and gaseous exchange in rats)
124. *Denisiewicz, R., Dziuk, E., Siekierzyński, M.*: Ocena czynności tarczycy u osób zatrudnionych w zasięgu promieniowania mikrofalowego. Polskie Arch. Med. Wewn. **45**, 19, 1970.
(Evaluation of thyroid function in persons profesionally exposed to microwaves)
125. *Djatchenko, N. A.*: Profilaktika funkcjonalnych naruszenii serdetchnososudistoi sistemy u operatorov RLS. Vojenno-Med. Zhurnal **9**, 45, 1970.
(Prophylaxis of functional cardiovascular disturbances in radar operators)
126. *Djatchenko, N. A.*: Izmenenie funkcii shtchytovidnoi zhelezy pri chronitcheskom vozdeistvii polja SVTCh. Gig. Truda i Profzabolevanii **7**, 51, 1970.
(Changes in thyroid function induced by chronic exposure to SHF fields)
127. *Dodge, C. H.*: Foreign Sci. Bull. **1**, 7, 1965, Library of Congress.
128. *Dodge, C. H.*: ATD Report P-65-68. Library of Congress, 1965, 93 pp.
129. *Dodge, C. H.*: ATD Report P-65-17. Library of Congress, 1965, 44 pp.
130. *Dodge, C. H.*: Clinical and hygienic aspects of exposure to electromagnetic fields. In [82], p. 140.
131. *Dodge, C. H., Kassel, S.*: ATD Report 66-133, Library of Congress, 1966, 33 pp.
132. *Dolatowski, A., Leńko, J., Mróz-Wasilewska, Z., Wochna, Z.*: Badania nad wpływem mikrofal aparatury radarowej na jądra i najądrza królików. Polski Przegląd Chirurgiczny **35**, 1221, 1963 (also Polish Med. J. **5**, 1156, 1964).

(Investigations on the influence of microwaves of radar equipment on testes and epididymis of rabbits)
133. *Dordević, Z.*: Zdravstveni pregledi radarista. Vojnosanitetski Pregled **7**, 460, 1966.
(Health examination of radar workers)
134. *Doyle, J. R., Smart, B. W.*: Stimulation of bone growth by short-wave diathermy. J. Bone Joint Surg. **15**, 45A, 1963.
135. *Drogitchina, E. A., Sadtchikova, M. N.*: Klinitcheskie sindromy pri vozdeistvii razlitchnych diapazonov radiotchastot. Gig. Truda i Profzabolevanii **1**, 17, 1965.
(Clinical syndromes induced by various radio-frequency-wave ranges)
136. *Drogitchina, E. A., Sadtchikova, M. N., Ginzburg, D. A., Tchulina, N. A.*: Nekotorye klinitcheskie projavlenija chronitcheskogo vozdeistvija santimetrovych voln. Gig. Truda i Profzabolevanija **1**, 28, 1962.
(Certain clinical symptoms of chronic influence of centimeter waves)
137. *Dryden, J., Jackson, W.*: Dielectric behaviour of methylpalmitate: evidence of resonance absorption. Nature **162**, 656, 1948.
138. *Duane, T. D., Hines, H. M.*: Experimental lenticular opacities produced by microwave irradiations. Arch. Phys. Med. **28**, 765, 1948.
139. *Duhamel, J.*: Effects biologiques des ondes courtes. Presse Med. **66**, 744, 1958.
(Biological effects of short waves)
140. *Dziuk, Z., Denisiewicz, M., Siekierzyński, M., Symonowicz, N.*: Krzywe przecukrzenia u osób narażonych na przewlekłe działanie promieniowania mikrofalowego. Lek. Wojsk. **46**, 884, 1970.
(Sugar curves in individual exposed to chronic microwave irradiation)
141. *Eastwood, E.*: Radar ornithology. Methuene Company Ltd. London, 1967.
142. *Edelwejn, Z.*: Próba oceny stanu czynnościowego synaps mózgowych królików poddanych przewlekłemu działaniu mikrofal. Acta Physiol. Polon. **19**, 897, 1968.
(An attempt at evaluation of the functional state of brain synapses in microwave-irradiated rabbits)
143. *Edelwejn, Z., Barański, S.*: Badania nad wpływem warunków promieniowania na układ nerwowy osób zatrudnionych w zasięgu działania mikrofal. Lek. Wojsk. **42**, 781, 1966.
(Investigations on the influence of irradiation conditions on the central nervous system in personnel professionally exposed to microwaves)
144. *Edelwejn, Z., Barański, S.*: Badania elektroencefalograficzne i morfologiczne wpływu promieniowania mikrofalowego na niektóre głębokie struktury ośrodkowego układu nerwowego u królików. Postępy Fiz. Medycznej **6**, 145, 1971.
(Morphologic and electroencephalographic investigations on the influence of microwaves on certain deep central nervous system structures in rabbits)
145. *Egan, W. G.*: Eye protection in radar fields. Elec. Eng. **2**, 126, 1957.
146. Electromagnetic radiation hazards. Ground Electronics Engineering-Installation Agency Standard. T. O. 31 Z-10-4. Aug. 1966.
147. *Elfring, G.*: Effects of the local application of heat on the physiology of the testes. Sahalan Kirjapaino Oy. Helsinki, 1950.
148. *Ely, T. S.*: Microwave death. J. Am. Med. Assoc. **217**, 1394, 1971.
149. *Ely, T. S., Goldman, D. E.*: Heat exchange characteristics of animals exposed to 10 cm microwaves. IRE Trans. Med. Electron. vol. PGME-4, p. 38, 1956.
150. *Ely, T. S., Goldman, D. E., Hearon, J. Z., Williams, R. B., Carpenter, H. M.*: Heating characteristics of laboratory animals exposed to ten-centimeter microwaves. U. S. Navy, Naval Medical Research Institute, Bethesda. Md. Research Report, Project NM 001.256. 13.02. 1957, vol. 15, p. 77. (Citation according to [353]).
151. *Ely, T. S., Goldman, D. E., Hearon, J. C.*: Heating characteristics of laboratory animals exposed to ten-centimeter microwaves. IEEE Trans. Biomed. Electron., vol. **ME-11**, 123, 1964.

152. *England, T., Sharpless, N.*: Dielectric properties of the human body in the microwave region of the spectrum. Nature **163**, 487, 1949.
153. *England, T.*: Dielectric properties of the human body for wavelengths 1-10 cm range. Nature **166**, 480, 1950.
154. *Engle, J. P., Herrick, J. F., Wakim, K. G., Grindlay, J. H., Krusen, F. H.*: The effect of microwaves on bone and bone marrow and adjacent tissues. Arch. Phys. Med. **31**, 453, 1950.
155. *Erdman, R. A.*: Microwave exposure of the human female pelvis during early pregnancy and prior to conception, case report. Am. J. Phys. Med. **38**, 219, 1959.
156. *Ermolajev, E. A.*: Zashtchita ljudei od voizdeistvii SVTCh izlutchenija. In [425], p. 189. (Protection of personnel against the influence of UHF radiation)
157. *Esman, L., Wise, C. S.*: Local effects of microwave radiation on tissues in the albino rat. Arch. Phys. Med. **31**, 502, 1950.
158. *Fajtelberg-Blank, V. R.*: Vlijanie vysokotchastotnych voln sentimetrovogo diapozona na vsasyvatielnuju dejatelnost sheludka i kishetchnika. Bjul. Eksper. Biol. Med. **57**, 45, 1964. (Effect of high-frequency microwaves on the absorptive activity of the stomach and intestine)
159. *Faitelberg-Blank, V. R.*: Vsasyvatielnoja dejatelnost zheludka i kishetchnika po vlijanin elektricheskogo polja UVTch. Fizjologitcheskij Zhurnal SSSR in J. M. Setchenova, **48**, 735, 1962. (Absorptive activity of stomach and intestine after exposure to electric UHF fields)
160. *Fleming, J. Jr., Pineo, L., Baus, R. Jr., McAfee, R.*: Microwave radiation in relation to biological systems and neural activity. In [427], p. 229.
161. *Follis, R. H.*: Am. J. Physiol. **147**, 281, 1946. (Citation according to [353]).
162. *Frey, A. H.*: Brain stem evoked responses associated with low intensity pulsed UHF energy. J. Appl. Physiol. **23**, 984, 1967.
163. *Frey, A. H.*: Behavioral biophysics. Psychol. Bull. **63**, 322, 1965.
164. *Frey, A. H.*: Biological function as influenced by low power modulated RF energy. IEEE Trans. Microwave Theory and Techniques, vol. MTT-19, 153, 1971.
165. *Frey, A. H.*: Effects of microwaves and radiofrequency energy on the central nervous system. In [82], p. 134.
166. *Frey, A. H.*: Auditory system response to radio-frequency energy. Aerospace Med. **32**, 1140, 1961.
167. *Frey, A. H.*: Human auditory system response to modulated electromagnetic energy. J. Appl. Physiol. **17**, 689, 1962.
168. *Frey, A. H.*: Some effects on human subjects of ultrahigh frequency radiation. Am. J. Med. Electron. **2**, 28, 1963.
169. *Frey, A. H., Fraser, A., Siefert, E., Brish, T.*: A coaxial pathway for recording from the cat brain stem during illumination with UHF energy. Physiol. Behav. **3**, 363, 1968.
170. *Frey, A. H., Siefert, E.*: Pulse modulated UHF energy illumination of the heart associated with change in heart rate. Life Sci. **7**, 505, 1968.
171. *Frey, A. H., Thornton, S.*: A restraint device for cats in a UHF electromagnetic energy field. Psychophysiology **63**, 381, 1966.
172. *Frolova, L. T.*: K gigienitcheskoi ocenke uslovii truda pri rabote s tokami sverchvysokoi tchastoty. Gig. Truda i Profzabolevanii **2**, 27, 1963. (Hygienic evaluation of work conditions in dealing with ultra-high-frequency currents)
173. *Galanin, H. F.*: Uslovia truda operatora na radiolokacjonnych ustanovkach i vozmozhnye puti obshtchego i zritelnogo utomlenija. Voienno-Medicinski Zhurnal **9**, 28, 1956. (Work conditions of radiolocation station operators and possible means of prophylaxis of general and visual fatigue)
173. *Gavrilova, O. E.*: O vlijanii SVTCh-polja na synaptitcheskuju peredatchu vozbuzhdenija

v vierchniem shejnom sympatitcheskom ganglii koshki. In: Voprosy biologitcheskogo diejstvija sverchvysokotchastotnogo (SVTCh) elektromagnitnogo polja. Leningrad, 1962, p. 13.
(The influence of SHF-field on synaptic transmission of excitation in the cranial neck sympathetic ganglion of the cat)

175. *Gelfan, I. A., Sadtchikova, M. N.*: Bielkovuje frakcji i gistamin krovi pri vozdeistvi SVTch i VTch. In [294], p. 46.
(Protein fractions and histamine of the blood after exposure to SHF and HF)

176. *Gembickii, E. W.*: Materiały k klinike chronitcheskogo vozdiestvia mikrovoln. Abstr. Tezisi Nautchnoj Konferencji. Leningrad, 1962.
(Materials on the treatment of chronic exposure to microwaves)

177. *Gersten, J. W., Wakim, K. G.*: The effects of microwave diathermy on the peripheral circulation and on tissue temperature in man. Arch. Phys. Med. **30**, 7, 1949.

178. *Gifford, E. C., Lazo, B. A.*: Radar target detection under white collimated lighting. Aerospace Med. **3**, 336, 1962.

179. *Glaser, Z. R., Heimer, G. M.*: Determination and elimination of hazardous microwave fields aboard naval ships. IEEE Trans. Microwave Theory Techniques, vol. **MTT-19**, 232, 1971.

180. *Gordon, E. W.*: Itogi nautchnogo soveshtchenija po voprosam dozimetrii i gigienitcheskoi ocenki elektromagnitnych polei pri rabote z generatorami vysokich i sverchvysokich tchastot. Gig. Sanitaria **20**, 56, 1955.
(Problems of scientific investigations on dosimetry and hygienic evaluation of electromagnetic fields during work with high- and super-high-frequency generators)

181. *Gordon, Z. V.*: Voprosy gigieny truda biologitcheskogo dejstvija radiovoln rezlitchnych dispozonov. Viestnik Akademii Medicinskich Nauk **119**, 42, 1964.
(Problems of industrial hygiene and biological effects of radiowaves of various frequencies)

182. *Gordon, Z. V.*: Gigienitcheskaja ocenka uslovii truda rabotajushtchych s generatorami SVTch. In [294], p. 22.
(Hygienic evaluation of working conditions of personnel servicing SHF generators)

183. *Gordon, Z. V.*: Vlijanie mikrovoln na uroven krovjanogo davlenija v eksperimentie na zhivotnych. O biologitcheskom diejstvii elektromagnitnych polei radiotchastot. Medgiz. Moscow, 1964, p. 57.
(The influence of microwaves on blood pressure levels in animal experiments)

184. *Gordon, Z. V.*: K voprosu o biologitcheskom deistvii sverchvysokich tchastot. Trudy Instituta Gig. Truda Profzabolevenii AMN SSSR **1**, 5, 1960.
(On the problem of the biological effects of superhigh frequencies)

185. *Gordon, Z. V.*: Voprosy gigieny truda i biologitcheskogo diejstvija elektromagnitnych polei sverchvysokich tchastot. Medicina. Moscow, 1966.
(Problems of industrial hygiene and the biological effects of electromagnetic super-high-frequency fields)

186. *Gordon, Z. V.*: Occupational health aspects of radio-frequency electromagnetic radiation. Proc. ILO-ENPI International Symposium on Ergonomics and Physical Environmental Factors. Rome, 1968. International Labor Office, Geneva, 1970.

187. *Gordon, Z. V., Lobanova, E. A.*: Temperaturnaja reakcija zhivotnych pri vozdeistvii SVTch. In [294], p. 59.
(Temperature response in animals exposed to SHF)

188. *Gordon, Z. V., Lobanova, E. A., Kicovskaja, I. A., et al.*: Issledovanie biologitcheskogo deistvija elektromagnitnych voln millimetrovogo diapozona. Bjul. Eksper. Biol. Med. **68**, 37, 1969.
(Investigations on biological effects of millimeter electromagnetic waves)

189. *Gordon, Z. V., Lobanova, E. A., Nikogosjan, S. V., Kicovskaja, I. A., Tolgskaja, M. S.*: K voprosu o niekotorych ossobiennostiach biologitcheskogo deistvija razlitchnych diapozonov mikrovoln. In Voprosy biologitcheskogo deistvija sverchvysokotchastotnogo (SVTch) elektromagnitnogo polia. Leningrad, 1962, p. 15.

(On the problem of certain peculiarities of biological effects of various microwave bands)
190. *Gordon, Z. V., Lobanova, E. A., Tolgskaja, M. S.*: Nekotoryje dannyje o deistvie santimetrovych voln. Gig. Sanitaria **12**, 16, 1955.
(Certain data on centimeter-wave effects)
191. *Gorodeckaja, S. F.*: Morfologitchne zmiany vnutriennych organov pri vplyvie na organizm santimetrovych chvyl. Fizjologitchnyi Zhurnal **8**, 390, 1962.
(Morphologic changes of internal organs during exposure to centimeter waves)
192. *Gorodeckaja, S. F.*: Vlijanie radiovoln 3-santimetrovogo diapozona na funkcionalnoe sostojanie kory nadpotchetchnikov. Fizjol. Zhurnal Akad. Nauk USSR **7**, 672, 1961.
(Influence of radiowaves in the 3-cm range on the functional state of the adrenal cortex)
193. *Gorodeckaja, S. F.*: Vlijanie SVTch-elektromagnitnogo polia na razmozhenie, sostav periferitcheskoi krovi, uslovnoreflektornuju dejatelnost i morfologiju vnutrennych organov belych myshei. In [195], p. 80.
(The influence of SHF electromagnetic field on fertility, peripheral blood picture, conditioned reflexes, and morphology of internal organs of white mice)
194. *Gorodeckaja, S. F.*: K voprosu o vlijanii radiovoln santimetrovogo diapozona na vysshuju nervnuju dejstelnost, organy krovetvorenija i razmnozhenija. Fizjol. Zhurnal A. N. SSSR **6**, 622, 1960.
(On the problems of influence of centimeter radio-waves on higher nervous function, blood-forming systems, and fertility)
195. *Gorodeckij, A. A.* (ed.): Biologitcheskoje deistvie ultrazvuka i zverchvysokotchastotnych elektromagnitnych kolebanii. Naukova dumka, Kiev, 1964.
(Biological effects of ultra-sound and ultra-high-frequency electromagnetic waves)
196. *Górski, S., Kwaśniewska-Błaszczyk, M., Mackiewicz, S., Ramlau, D.*: Ocena izotopowa działania mikrofal na krążenie włośniczkowe w mięśniach kończyn. Pol. Tyg. Lek. **22**, 940, 1967.
(Izotopic evaluation of the influence of microwaves on capillary circulation in the muscles of extremities)
197. *Granberry, W. M., Janes, J. M.*: The lack of effect of microwave diathermy on rate of growth of bone of the growing dog. J. Bone Joint Surg. **45A**, 773, 1963.
198. *Grebieshnikova, A. M.*: K voprosu o vlijanii SVTch-polia decimetrovogo i metrovogo diapozonov voln na dvigatelno-evakuatornuju funkciju zheludotchno-kishetchnego trakta sobak i morskich svinok. In: Voprosy biologitcheskogo deistvija sverchvysokotchastotnogo (SVTch) elektromagnitnogo polia. Tezisy Nautchnoi Konferencji (Proc. Scient. Conference). Leningrad, 1962, p. 17.
(On the influence of UHF fields in the centimeter- and meter-wave range on the mobility and evacuation of the stomach and intestines in dogs and guinea pigs)
199. *Greenfield, J. G.*: Metachronic leuko-encephalopathy. In: Neuropathology. Charles C Thomas Springfield, Ill. 1959, p. 465.
200. *Griffin, J. L.*: Orientation of human and avian erythrocytes in radiofrequency fields. Exp. Cell. Res. **61**, 113, 1970.
201. *Griffin, J. L., Ferris, C. D.*: Pearl chain formation across radiofrequency fields. Nature **226**, 152, 1970.
202. *Gruszecki, L.*: Badania doświadczalne nad wpływem mikrofal wysyłanych przez nadajniki radarowe na ustrój ludzki i zwierzęcy. Biul. Wojskowej Akad. Med. **18**, suppl. II, 1964.
(Experimental investigations on the influence of radar waves on the human and animal organism)
203. *Gruszecki, L.*: Badania doświadczalne nad wpływem mikrofal wysyłanych przez nadajniki radarowe na ustrój ludzki i zwierzęcy. Biul. Wojskowej Akad. Med. **5**, 61, 1962.

(Experimental investigations on the influence of radar microwaves on the human and animal organism)
204. *Guan, K. C.*: Hazards of microwave radiations: A review. Ind. Med. Surg. **29**, 315, 1960.
205. *Gunn, S. A., Gould, T. C., Anderson, W. A. D.*: The effect of microwave radiation (24,000 Mc) on the male endocrine system of the rat. Ind. Med. Surg. **30**, 295, 1961 (see also in [82], p. 99).
206. *Gunn, S. A., Gould, T. C., Anderson, W. A. D.*: The effects of microwave radiation on morphology and function of rat testis. Lab. Invest. **10**, 303, 1961.
207. *Guriev, V. N.*: Sindromy vyzivanyje chronitcheskim vozdeistviem elektromagnitnych voln. Abstr. Tezisy Nautchnoi Konferencji. Leningrad, 1962.
(Syndromes induced by chronic electromagnetic wave irradiation)
208. *Guriev, V. N.*: Nekotorye voprosy adaptacji ludei k SVTch vozdeistviju v proizvodstvennych uslovijach. Abstr. Tezisy Nautchnoi Konferencji. Leningrad, 1962.
(Certain problems of adaptation of man to SHF exposure in industrial environment)
209. *Guy, A. W.*: Measurement of absorbed electromagnetic power density in biological systems. Abstr. Microwave Symposium. Boulder, Col. 1972, p. 69.
210. *Guy, A. W.*: Electromagnetic fields and relative heating patterns due to a rectangular aperture source in direct contact with bilayered biological tissue. IEEE Trans. Microwave Theory Techniques, vol. **MTT-19**, 214, 1971.
211. *Guy, A. W.*: Analyses of electromagnetic fields induced in biological tissues by thermographic studies on equivalent phantom models. IEEE Trans. Microwave Theory Techniques, vol. **MTT-19**, 205, 1971.
212. *Guy, A. W., Lehmann, J. F.*: On the determination of an optimum microwave diathermy frequency for a direct contact applicator. IEEE Trans. Biomed. Eng., vol. **BME-13**, 76, 1966.
213. *Gvozdikova, Z. M., Ananjev, V. M., Zienina, I. N., Zak, V. J.*: O tchustvitielnosti centralnoi nervnoi sistemy krolokov k nieprerivnomu elektromagnitnomu polju sverchvysokich tchastot. Bjul. Eksper. Biol. Med. **58**, 63, 1964.
(On the sensitivity of the central nervous system of rabbits to continous exposure to electromagnetic super-high-frequency fields)
214. *Haduch, S., Barański, S., Czerski, P.*: Wpływ promieniowania mikrofalowego na ustrój ludzki. Acta Physiol. Polon. **11**, 717, 1960.
(The influence of microwaves on the human organims)
215. *Haduch, S., Barański, S., Czerski, P.*: Badania nad wpływem pola elektromagnetycznego wysokiej częstotliwości na ustrój ludzki. Doniesienie I — krew obwodowa. Lek. Wojsk. **2**, 119, 1966.
(Investigations on the influence of high-frequency electromagnetic fields on the human organism. I. Peripheral blood picture)
216. *Haduch, S., Czerski, P., Barański, S.*: Biologiczne działanie fal centymetrowych i decymetrowych. Lek. Wojsk. **36**, 792, 1960.
(Biological effects of centimeter and decimeter waves)
217. *Harvey, A. F.*: Industrial, biological and medical aspects of microwave radiation. Proc. IEE 107, 557, 1960.
218. *Hauswirth, C., Kraemer, F.*: Über die Wirkung von Mikrowellen auf das vegetative System. Wien. Med. Wochenschr. **108**, 172, 1958.
(On the influence of microwaves on the neurovegetative system)
219. *Healer, J.*: Review of studies of people occupationally exposed to radio-frequency radiations. In [82], p. 90.
220. *Hearn, G. E.*: Effects of UHF radio fields on visual acuity and critical flicker fusion in the albino rat. Ph. D. dissertation. Baylor University, Waco, Texas, 1965.
221. *Heller, J. H.*: Cellular effects of microwave radiation. In [82], p. 116.

222. *Heller, J. H., Mickey, G. H.*: Nonthermal effects of radiofrequency in biologic systems. Digest International Conference on Medical Electronics. New York, 1961, p. 152.
223. *Heller, J. H., Teixeira-Pinto, A. A.*: A new physical method of creating chromosomal aberrations. Nature **183**, 905, 1959.
224. *Herrick, J. F.*: Pearl chain formation. Proc. Second Tri-Serv. Conf. Biol. Effects Microwave Energy. Rome, N. Y. 1958, p. 83.
225. *Herrick, J. F., Krusen, F. H.*: Problems which are challenging investigators in medicine. IRE Trans. Med. Electron., vol. **PGME-4**, 10, 1956.
226. *Herrick, J. F., Krusen, H. P.*: Certain physiologic and pathologic effects of microwaves. Elec. Eng. **70**, 239, 1953.
227. *Herrington, L. P.*: The heat regulation of small laboratory animals at various environmental temperatures. Am. J. Physiol. **129**, 123, 1940.
228. *Hines, H. M., Randall, J. E.*: Possible industrial hazards in the use of microwave radiation. Elec. Eng. **71**, 879, 1952.
229. *Hirsch, F. G., Parker, J. T.*: Bilateral lenticular opacities occurring in a technician operating a microwave generator. Arch. Ind. Hyg. **6**, 512, 1952.
230. *Ho, H. S., Guy, A. M., Sigelman, R. A., Lehmann, J. F.*: Microwave heating of simulated human limbs by aperture sources. IEEE Trans. Microwave Theory Techniques, vol. **MTT-19**, 224, 1971.
231. *Hofman, D.*: Experimentelle Untersuchungen über die ^{32}P und ^{35}S. Inkorporation in den Genitalorganen der Ratt und den Einfluss der Kurzwellenbestrahlung. Zbl. Gynaekol. **91**, 593, 1969.
(Experimental studies on the incorporation of ^{32}P and ^{35}S in the genital organs of the rat and its dependence on shortwave irradiation)
232. *Hoeft, L. O.*: Microwave heating: a study of the critical exposure variables for man and experimental animals. Aerospace Med. **36**, 621, 1965.
233. *Hollander, A. (ed)*: Radiation Biology. McGraw-Hill Book Company, New York, Toronto, London. Vol. I, 1954; vol. II, 1955; vol. III, 1956.
234. *Horai, H.*: Biological effects of microwave radiation. I. Alteration of the rectal temperature of the mice during the microwave radiation and microscopic findings of destroyed cases. Nippon Acta Radiol. **22**, 173, 1962 (original title and text in Japanese with English summary).
235. *Hornowski, J., Marks, E., Chmurko, E., Pannert, L.*: Z badań nad wpływem chorobotwórczym mikrofal u ludzi. Med. Pracy 17, 213, 1966.
(On the pathogenic influence of microwaves in man)
236. *Hornowski, J, Marks, E., Chmurko, E., Pannert, L.*: Badania kliniczne osób zatrudnionych przy mikrofalach. Med. Lotnicza **18**, 39, 1965.
(Clinical examination of personnel professionally exposed to microwaves)
237. *Howland, J. W., Michaelson, S.*: Studies on the biological effects of microwave irradiation o the dog and rabbit. In [518], p. 191.
238. *Howland, J. W., Michaelson, S., Thomson, E., Mermagen, H.*: The effect of microwaves on the response to ionising radiation. Rep. RADS-TDR 62-102, Univ. Rochester, Rochester, N. Y., 1962.
239. *Howland, J. W., Thomson, R. A. E., Michaelson, S. M.*: Biomedical aspects of microwave irradiation of mammals. In [427], p. 261.
240. *Hutt, B. K., Moore, J., Coloma, P. C., Horvath, S. M.*: The influence of microwave irradiation on bone temperature in dog and man. Am. J. Phys. Med. **31**, 422, 1952.
241. *Iberall, A. S.*: Human body as an inconstant heat source and its relation to determination of clothes insulation. In [518], pp. 137 and 147.
242. *Illinger, K. H.*: Molecular mechanisms for microwave absorption in biological systems. In [82], p. 112.

243. *Imig, C. J., Thomson, J. D., Hines, H. M.*: Testicular degeneration as a result of microwave irradiation. Proc. Soc. Exptl. Biol. Med. **69**, 382, 1948.
244. *Ivanov, A. J.*: Izmenenija fagocitarnoi aktivnosti i podviznosti neitrofilov pod vlijanijem SVTch polja. Abstr. Tezisy Nautchn. Konf., Leningrad, 1962, p. 34.
(Changes in phagocytic activity and mobility of neutrophils under the influence of UHF fields)
245. *Jackson, W.*: Dielectric behaviour of methylpalmitate. Nature **164**, 486, 1949.
246. *Jacobson, B. S., Prausnitz, S. B., Süsskind, C.*: Investigation of thermal balance in mammals by means of microwave radiation. IRE Trans. Med. Electron., vol. **ME-6**, 66, 1959.
247. *Jakovleva, M. I., Shliafer, T. P., Cvetkova, I. P.*: K voprosu ob uslovnych serdetchnych refleksach, funkcionalnom i morfologitcheskom sostojanii korovych neironov pri deistvii elektromagnitnych polei sverchvysokich tchastot. Zhurnal Vysshei Nervnoi Dejatelnosti **18**, 973, 1968.
(On the problems of conditioned cardiac reflexes, functional and morphologic state of cortical neurons during exposure to super-high-frequency electromagnetic fields)
248. *Janes, D. E., Leach, W. M., Mills, W. A., Moore, R. T., Shore, M. L.*: Effects of 2450-MHz microwaves on protein synthesis and on chromosomes in Chinese hamsters. Non-ionizing Radiation **1**, 125, 1969.
249. *Jankowiak, J., Majewski, C.*: Einfluss der Mikrowellen auf die Rattenleber auf Grund histologischer und histochemischer Untersuchungen. Hochfrequenz Therapie **9**, 3, 1959.
(The influence of microwaves on the liver of rats, histologic and histochemical studies)
250. *Janiszewski, S., Szymańczyk, L.*: Badania nad wpływem mikrofal na narząd wzroku. Med. Lotnicza **2**, 1969, p. 81.
(Investigations on the influence of microwaves on the eye)
251. *Jaski, T., Süsskind, C.*: Electromagnetic radiation as a tool in life sciences. Science **133**, 443, 1961.
252. *Jasser, S., Brągiel, I., Czerski, P., Polubiec, A.*: Badania nad układem properdyny i innych czynników naturalnej odporności u osób zatrudnionych w zasięgu mikrofal. Lek. Wojsk. **7**, 517, 1963.
(Investigations on the properdine system and other factors of natural immunity in persons professionally exposed to microwaves)
253. *Johnson, C. C., Guy, A. W.*: Non-ionizing electromagnetic wave effects in biological materials and systems. Proc. IEEE **60**, 692, 1972.
254. *Joly, R., Plurien, G., Dromet, J., Servantie, B.*: Effect biologiques et physiopathologiques éventuels des rayonnements électromagnetiques UHF des "aériens-radars". Rev. Corps de Santé des Armées **10**, 239, 1969.
(Biological and eventual physiopathological effects of UHF electromagnetic radar radiations)
255. *Ju-Tchzhin*: Izmenenie mezhutotchnogo vieshtchestva settchatki pod vlijaniem lutchystoi energji. Vestnik Oflatmol. **4**, 5, 1959.
(Changes of intercellular substance of the retina under the influence of radiant energy)
256. *Justesen, D. R., King, N. W.*: Behavioral effects of low level microwave irradiation in the closed space situation. In [82], p. 154.
257. *Kadaravek, F.*: Termoregulacni deje pri aplikaci mikrovln. Fysiat. Reum. Vestn. **48**, 112, 1970.
(Thermoregulatory response to microwaves)
258. *Kalant, H.*: Physiological hazards of microwave radiation. A survey of literature. Can. Med. Assoc. J. **81**, 575, 1959.
259. *Kamenskii, J. I.*: Deistvie mikrovoln na funkcionalnoje sostojanie nerva. Biofizyka **9**, 695, 1964.
(The influence of microwaves on the functional state of the nerve fiber)

260. *Kamat, G. P., Janes, D. E.*: Studies on the effect of 2450 MHz microwaves on human immunoglobulin G. In [82], p. 104.
261. *Kaplan, I. T., Metlay, W., Zaret, M. M., Birenbaum, L., Rosenthal, S. W.*: Absence of heartrate effects in rabbits during low-level microwave irradiation. IEEE Trans. Microwave Theory Techniques, vol. **MTT-19**, 168, 1971.
262. *Kerova, N. I.*: Vlijanie SVTCh elektromagnitnogo polja na aktivnost polinuklear i soderrhanie nukleinovych kislot. In [195], p. 108.
 (Influence of SHF electromagnetic fields on polynuclease activity and nucleic acid content)
263. *Kicovskaja, I. A.*: Vlijanie santimetrovych voln rozlitchnych intensivnosti na krov i krovetvornye organy belych krys. Gig. Truda Profzabolevanii **6**, 14, 1964.
 (Influence of centimeter waves at various power densities on the blood and blood-forming organs of white rats)
264. *Kicovskaja, I. A.*: Sravnitelnaja ocenka vozdeistvija mikrovoln raznych diapozonoc na nervnuju sistemu krys tchustvitelnych k zvukovomu razdrazheniju. In: O biologitcheskom deistvii elektromagnitnych polei radiotchastot. Medgiz. Moscow, 1964, p. 39.
 (Comparative evaluation of the influence of various microwave bands on the nervous system of rats sensitive to acoustic stimulation)
265. *Kicovskaja, I. A.*: Issledovanii vzaimootnoszenii miezhdu osnovnymi nervnymi procesami u krys pri vozdiejstvii SVTch. In: O biologitcheskom vozdiejstvii sverchrysokich tchastot. Moscow, 1960, p. 75.
 (Investigations on the relationship between basic nervous processes (functional phenomena) in rats exposed to SHF)
266. *Kierebiński, Cz.*: Porównanie strawności białka mięsa nagrzewanego mikrofalami (2450 MHz) i sposobami tradycyjnymi. Lek. Wojsk. **44**, 673, 1968.
 (Comparison of digestibility of meat cooked in a microwave oven (2450 MHz) and by traditional methods)
267. *Kierebiński, Cz.*: Przygotowanie potraw za pomocą mikrofal i ocena użytkowa urządzenia mikrofalowego. Lek. Wojsk. **44**, 980, 1968.
 (Cooking with microwaves and the evaluation of microwave equipment utility)
268. *Kidd, J. S.*: A summary of research methods, operator characteristics and system design specification on the study of a simulated radar air traffic control system. Aerospace Med. **31**, 90, 1961.
269. *King, N. W.*: The effects of low level microwave irradiation upon reflexive, operation and discrimination behaviors of the rat. Ph. D. dissertation. Univ. Kansas, Lawrence, Kans., 1969.
270. *King, N. W., Justesen, D. R., Clarke, R. L.*: Behavioral sensitivity to microwave irradiation. Science **172**, 398, 1971.
270a. *Klimkova-Deutschova, E., Macek, Z., Roth, E.*: Electroencephalographic study of neuroses and pseudoneuroses with particular emphasis on the electroencephalographic signs of reduced vigilance. Čas. lek. čs. **98**, 1213, 1959.
270b. *Klimkova-Deutschova, E.*: Strahlungseinfluss auf das Nerven system. Arch. Gewerbepathol. Generbehyg. **16**, 72, 1957; also **20**, 1, 1963.
 (The effect of radiation on the nervous system)
271. *Knepton, J. C. Jr., Beischer, D. E.*: Effects of very high magnetic fields on the electroencephalogram of the squirrel monkey. Aerospace Med. **37**, 325, 1966.
272. *Knoppe, K. G.*: Parametry polei SVTCH opredeljajushtchie gigienitcheskuju ocenku uslovii truda i zadatchi ich izmerenija. In [294], p. 11.
 (The parameters of SHF fields that influence the hygienic evaluation of working conditions and the purpose of their measurements)

273. *Kolesnik, T. A.*: Klinika chronitcheskogo vozdiejstvija elektromagnitnych voln swierchvysokoj tchastoty. V. M. A. Leningrad, 1961.
(The clinical picture of chronic exposure to super-high-frequency electromagnetic waves)
274. *Kolesnik, T. A., Małyszew, W. M., Murassow, B. F.*: O naruszeniach endokrinnoi sistemy pri chronitcheskom vozdieistvii sverch vysokich tchastot polia. Voenno-Med. Żurnal **7**, 39, 1967.
(Endocrine disturbances induced by chronic exposure to super-high-frequency fields)
275. *Kołakowski, Z.*: Zmiany chorobowe występujące u osobników zdrowych zatrudnionych w zasięgu promieniowania mikrofalowego. Lek. Wojsk. **47**, 309, 1971.
(Pathological changes in healthy personnel exposed professionally to microwaves)
276. *Kondra, P. A., Smith, W. K., Hodgson, G. C., Bragg, D. B., Gavora, J., Hamid, M. A. K., Bonlanger, R. J.*: Growth and reproduction of chickens subjected to microwave radiation. Can. J. Animal Sci. **50**, 639, 1970.
277. *Kondrat, J.*: Wpływ fal radaru na gojenie ran skóry u szczurów. Thesis, Medical Academy, Gdańsk, 1966. (The influence of radar waves on the healing of skin wounds)
278. *Kondrat, J., Meyer, J.*: Gojenie ran chirurgicznych skóry szczurów poddanych uprzednio ekspozycji pola elektromagnetycznego wysokiej częstotliwości. Lek. Wojsk. **45**, 508, 1967.
(Healing of surgical skin wounds in rats after prior exposure to electromagnetic high-frequency fields)
279. *Kondrat, J., Meyer, J.*: Gojenie ran chirurgicznych skóry szczurów poddanych wtórnie ekspozycji pola elektromagnetycznego wysokiej częstotliwości. Lek. Wojsk. **43**, 605, 1967.
(Healing of surgical skin wounds in rats exposed to high-frequency electromagnetic fields after wounding)
280. *Korbel, S. F.*: Behavioral effects of low intensity UHF radiation. In [82], p. 180.
281. *Korbel, S. F., Fine, H. L.*: Effects of low intensity UHF radio fields as a function of frequency. Psychonomic Sci. **9**, 527, 1967.
282. *Korbel, S. F., Thompson, W. D.*: Behavior effects of stimulation by UHF radio fields. Psychol. Rept. **77**, 595, 1965.
283. *Korytowski, E., Gruszecki, Z.*: Obserwacje nad wpływem krótkich fal elektromagnetycznych w stacjach radarowych na ustrój ludzki. Rocznik Służby Zdrowia Mar. Woj. (Gdynia), 1960, p. 139.
(Observations on the influence of short electromagnetic radar waves on the human organism)
284. *Kowatch, P. J.*: Trudy VMA im. S. M. Kirova, **166**, 160, 1966 (cited according to [425]).
285. *Kozłowski, B.*: Zmiany oczne u pracowników narażonych na działanie fal elektromagnetycznych wielkiej częstotliwości. Med. Pracy **14**, 487, 1963.
(Eye changes in personnel professionally exposed to very high frequency electromagnetic waves)
286. *Krusen, F. H., Herrick, J. F.*: Microkymatotherapy: Preliminary report on experimental studies of the heating effects of microwaves (radar) in living tissues. Proc. Staff Meetings Mayo Clinic, **22**, 209, 1949 (May).
287. *Kuchling, H.*: Physik. VEB Fachbuchverlag. Leipzig, 1968. (Physics).
288. *Kucia, H. R.*: Szkodliwe oddziaływanie wolnozmiennych pól magnetycznych na zdrowie pracowników. Ochrona Pracy **26**, 13, 1972.
(Injurious effects of low-frequency magnetic field in industrial workers)
289. *Kucia, H.*: Założenia do określenia bezpiecznych dawek promieniowania elektromagnetycznego z zakresu 0,1–300 MHz. Med. Lotnicza 39, 1972, p. 141.
(Premises for the determination of safe exposure limits to electromagnetic radiation in the 0.1–300 MHz range)
290. *Kudrjashova, V. A., Ilyina, S. A., Faleev, A. S., Gajdul, V. I., Dementienko, V. V.*: Issledovanie rozonansnogo vozdeistvija voln millimetrogo diapozona na gemoglobin. Abstr. symp. gig.

truda i biologitcheskoe deistvie elektromagnitnych voln radiotchastot. Moskva, 1972, p. 64. (Resonance millimeter electromagnetic wave radiation effect on hemoglobin)

291. *Kulakova, V. V.*: Vlijanie mikrovoln santimetrovogo i decimetrovogo diapozonov na obshtchie i specializirovannye formy appetita u zhirotnych. In O biologitcheskom deistvii elektromagnitnych polei radiotchastot. Moscow, 1964, p. 70.
(The influence of centimeter and decimeter waves on general and specialized forms of appetite in animals)

292. *Kulikovskaja, E. L.*: Elektromagnitnye polja sverchwysokich tchastot palubach gruzonych sudov. Gig. Truda i Profzabolevanii **2**, 24, 1963.
(Ultra-high-frequency electromagnetic waves on board merchant ships)

292a. *Kurz, H., Einaugler, R. B.*: Cataract secondary to microwave radiation. Am. J. Ophthalmol. **66**, 866, 1968.

293. *Lance, A. L.*: Introduction to microwave theory and measurements. McGraw-Hill Book Company, New York, 1964.

294. *Letavet, A. A., Gordon, Z. V.* (eds.): O biologitcheskom vozdeistvii sverchvysokich tchastot. Vol. I, Izd. Akad. Med. Nauk. SSSR. Moscow, 1960.
(English translation: "The biological action of ultrahigh frequencies". Office of Technical Services, U. S. Department of Commerce, Washington, D. C., 1962. Joint Publications. Research Service Rept. JPRS-12471; not always correct)

295. *De Lateur, B. J., Lehmann, J. F., Stonebridge, J. B., Warren, C. G., Guy, A. W.*: Muscle heating in human subjects with 915 MHz microwave contact applicator. Arch. Phys. Med. **51**, 147, 1970.

296. *Lawrence, J. C.*: Effect of pulsed microwaves at X-band frequency on skin metabolism. Nonionizing Radiation **1**, 80, 1969.

297. *Lawrence, J. C.*: Effect of microwaves at X-band on guinea pig skin to tissue culture. 1. Microwave apparatus for exposing tissue and the effect on skin respiratory. Brit. J. Ind. Med. **25**, 223, 1968.

298. *Leary, F.*: Researching microwave health hazards. Electronics **32**, 49, 1959.

299. *Lehmann, J. F.*: Comparative study of the efficiency of short wave microwave and ultrasonic diathermy in heating the hipjoint. Arch. Phys. Med. **40**, 510, 1959.

300. *Lehmann, J. F., Guy, A. W., Johnson, V. C.*: The comparison of relative heating in tissues by microwave about frequency 2450 and 900 Mc. Arch. Phys. Med. **43**, 69, 1962.

301. *Lehmann, J. F., Johnston, V. C., McMillan, J. A., Silverman, D. R., Brunner, G. D., Rathbun, L. A.*: Comparison of deep heating by microwave at frequencies of 2450 and 900 megacycles. Arch. Phys. Med. **46**, 307, 1965.

302. *Lehmann, J. F., McMillan, J. A., Brunner, G. D., Silverman, D. R., Johnston, K. C.*: Modification of heating patterns produced by microwaves at the frequencies of 2450 and 900 Mc by physiologic factors in the human. Arch. Phys. Med. **45**, 555, 1964.

303. *Lehmann, J., McMillan, J. A., Brunner, G. D., Guy, A. W.*: A comparative evaluation of temperature distributions produced by microwaves at 2456 and 900 megacycles in geometrically complex specimens. Arch. Phys. Med. **43**, 502, 1962.

304. *Leites, F., Skurichina, L. A.*: Vlijanie mikrovoln na gormonalnuju aktivnost kory nadpotchetchnikov. Bjul. Eksper. Biol. Med. **52**, 47, 1961.
(Influence of microwaves on hormonal activity of the adrenal cortex)

305. *Leńko, J., Dolatowski, A., Gruszecki, L., Klajman, S., Januszkiewicz, L.*: Wpływ 10 cm fal radaru na poziom 17-ketosterydów i 17-hydroksykortykosterydów w moczu królików. Przegląd Lekarski **22** (ser. II) 296, 1966.
(Influence of 10-cm radar waves on 17 KS and 17 HKS level in rabbit urine)

306. *Leńko, J., Waniewski, E., Wochna, Z., Dolatowski, A.*: Badania nad wpływem mikrofal aparatury radarowej na jądra królików. Biul. Wojsk. Akad. Med. **7**, 18, 1965.

(Investigations on the influence of microwaves emitted by radar equipment on rabbit testes)
307. *Libierman, E. A., Vajnstnajg, N. N., Cofina, L. M.*: K voprosu o deistvii postajannogo polja na porog vozbuzhdenija izolirovanego nierva ljagushki. Biofizika **4**, 505, 1959.
(On the problem of the influence of constant fields on the threshold of stimulation of the isolated frog nerve)
308. *Lidman, B. J., Cohn, C.*: Effect of radar emanations on the hematopoietic system. Air Surgeon's Bull. **2**, 448, 1945.
309. *Linke, C. A., Lounsberg, W., Goldschmidt, W.*: Effects of microwaves on normal tissues. J. Urol. **88**, 303, 1962.
310. *Lipowczan, A.*: Ilościowa ocena przepływowych pól elektromagnetycznych. Biul. Naukowy IMP w Przemyśle Węgl. i Hutn. 12/13, 91, 1965.
(Quantitation of industrial electromagnetic fields)
311. *Levitina, N. A.*: Issledovanie neteplovogo deistvija mikrovoln na ritm serdetchnoi dejatelnosti. Autoref. doctor's thesis. Moscow, 1966.
(Investigations on nonthermal effects on the cardiac rhythm)
312. *Levitina, N. A.*: Deistvie mikrovoln na ritm serdca krolika pri oblutchenii lokalnych utchastkov tela. Bjul. Eksper. Biol. Med. **65**, (7), 67, 1964.
(Influence of microwaves on the cardiac rhythm in rabbits during irradiation of limited body areas)
313. *Levitina, N. N.*: Issledovanie neteplovogo deistvija mikrovoln na ritm serdetchnych sokrashtchemii u ljagushki. Bjul. Eksper. Biol. Med. **67**, 64, 1966.
(Investigations of nonthermal effects of microwaves on the heart-beat rhythm in frogs)
314. *Livshic, N. A.*: Rol nervnoi sistemy w reakcijach organizma na deistvije elektromagnitnogo polja ultravysokoi tchastoty. Biofizika **2**, 378, 1957.
(The role of the nervous system in the responses of the organism to electromagnetic super-high-frequency fields)
315. *Livshic, N. N.*: Deistvije polja UVTch na funkcji nervnoi sistemy. Biofizika **3**, 426, 1958.
(Effects of UHF fields on the nervous system function)
316. *Livshic, N. N*: Uslovnoreflektornaja dejatelnost sobak pri lokalnych vozdeistvijach polem UVTch na nekotoryje zony kory bolshich polusharii. Biofizika **2**, 197, 1957.
(Conditioned reflexes in dogs during local exposure to UHF of certain cortical zones)
317. *Livshic, N. N.*: Uslovnoreflektornaja dejatelnost sobak pri vozdeistvii polem UVTch na oblast mozdzheka. Dokl. Akad. Nauk SSSR **112**, 145, 1957.
(Conditioned reflexes in dogs during exposure of the cerebellar region to UHF fields)
318. *Lobanova, E. A.*: Vyzhivajemost i rozvitie zhivotnych pri raznoi intensivnosti i dlitelnosti vozdeistvii SVTch. In [294], p. 61.
(Survival and development of animals during exposure to SHF for various periods and at various intensities)
319. *Lobanova, E. A.*: Izmenenija uslovnoreflektornoi dejatelnosti zhivotnych pri vozdeistvii mikrovoln razlitchnych tchastotnych diapozonov. In O biologitcheskom deistvii elektromagnitnych polei radiotchastot. Moscow, 1964, p. 13.
(Changes in conditioned reflexes in animals exposed to various microwave bands)
320. *Lobanova, E. A.*: Izmenija uslovnoreflektornoi dejatelnosti u zhivotnych (krys i krolikov) pri chronitcheskom vozdeistvii santimetrovych voln. In: Gigiena truda i biologitcheskoe deistvie elektromagnitnych voln radiotchastot. Tezisy dokladov. Moscow, 1959, p. 46.
(Changes in conditioned reflexes in animals (rats and rabbits) exposed to chronic centimeterwave irradiation)
321. *Lobanova, E. A., Tolgskaja, M. S.*: Izmenenije vysshei nervnoi dejatelnosti i mezhneironnych svjazei w kore golovnego mozga zhivotnych pri vozdeistvii SVTch. In [294], p. 69.

(Changes in higher nervous function and neuronal connections in the cerebral cortex of animals exposed to SHF)
322. *Lysina, G. C.*: Izmienienija morfołogitcheskogo sostava krovi pri vozdejstvii sverchvysokich tchastot. Gig. Sanitaria **30**, 95, 1965.
(Peripheral blood picture changes on exposure to superhigh frequencies)
323. *Łętowski, A.*: Badania doświadczalne nad wpływem aparatury radarowej na ustrój szczurów ze szczególnym uwzględnieniem narządów płciowych żeńskich. Ginekologia Polska, suppl. 7, 1966, p. 51. Biul. Wojsk. Akad. Med., suppl. II, 1967.
(Experimental investigations on the influence of radar equipment on rats in particular on female genital organs)
324. *Łętowski, A., Bartoszewicz, T., Lankienicki, A.*: Próba oceny metodą cytohormonalną i histochemiczną estrogenów u szczurzyc ciężarnych napromienianych mikrofalami. Lek. Wojsk. **47**, 551, 1971.
(Attempts at evaluation of endogenous estrogens in pregnant rats following microwave irradiation using cytohormonal and histochemical methods)
325. *Majewska, K.*: Badania nad wpływem mikrofal na narząd wzroku. Klinika Oczna **32**, 323, 1968.
(Investigations on the influence of microwaves on the eye)
326. *Majewski, C., Jankowiak, J., Głowacki, R.*: Influence des micro-ondes sur les organes interieurs des cobayes d'aprés des etudes histologiques et histochemiques. Minerva Medicophysica **7**, 8, 1971.
(Influence of microwaves on internal organs of guinea pigs, histologic and histochemical studies)
327. *Malachov, A. N. et al.*: Elektromagnitnoe pole SVTch kak signelnyi faktor v oboronitelnom uslovnom reflekse belych myshei. In: Materialy K 3 povolzhskoi konferencii fiziologov, biochimikov i farmakologov. Gorkii, 1963, p. 310 (citation according to [438]).
(Electromagnetic SHF field as a signal for conditioned defense reflex in white mice)
328. *Manczarski, S.*: Początek badań w Polsce biologicznego działania pól elektromagnetycznych i dalsze perspektywy. Med. Lotnicza 39, 1972, p. 7.
(The beginings of investigations on biological effects of electromagnetic fields in Poland and further perspectives)
329. *Manczarski, S.*: Oddziaływanie pola wielkiej częstotliwości na radiostacjach na organizm ludzki. Instytut Łączności. Warsaw, 1955.
(Effects of high-frequency field on the human organism in radio stations)
330. *Manczarski, S., Kucia, H.*: Ochrona pracowników przed szkodliwymi polami elektromagnetycznymi wielkiej częstotliwości. Ochrona Pracy **23**, 7–8, 24, 1969.
(Personnel protection against high-frequency field hazards)
331. *Manczarski, S., Mikke, D.*: Badania szkodliwości pól elektromagnetycznych wielkiej częstotliwości. Pomiary natężeń pól oraz środki zaradcze. Instytut Łączności. Warsaw, 1961.
(Investigations on electromagnetic high-frequency field hazards. Measurements and protection)
332. *Marha, K.*: Maximum admissible values of HF and UHF electromagnetic radiation at work places in Czechoslovakia. In [82], p. 188.
333. *Marha, K.*: Microwave safety standards in eastern Europe. IEEE Trans. Microwave Theory Techniques, vol. **MTT-19**, 165, 1971.
334. *Marha, K.*: Personal communication, 1972.
335. *Marks, E., Hornowski, J.*: Obserwacje kliniczne wpływu mikrofal na układ nerwowy. Neurol. i Neuroch. Polska **2**, 25, 1968.
(Clinical observations on the influence of microwaves on the nervous system)
336. *Marha, K., Musil, J., Tuha, H.*: Elektromagneticke pole a zivotni prostredi. Statni zdrovotnicke nakladatelstvi. Praha, 1968.

(English translation: Electromagnetic fields and the life environment. San Francisco Press, San Francisco, 1971)
337. *Martner, S.*: Personal communication, 1972.
338. *Martner, S.*: Occupational health in microwave industry. Abstr. XVII International Congress on Occupational Health. Buenos Aires, 1972, p. 116.
339. *Mazurkiewicz, J., Mielczarek, H., Zalejski, S., Siekierzyński, M.*: Zmiany w układzie nerwowym osób zatrudnionych w zasięgu mikrofal. Lek. Wojsk. **42**, 9, 1966.
(Central nervous system changes in personnel professionally exposed to microwaves)
340. *McAfee, R. D.*: Neurophysiological effects of microwave irradiation. Proc. Third Tri-Serv. Conf. on the Biological Effect of Microwave Radiating Equipment, 1959.
341. *McAfee, R. D.*: Physiological effects of thermal and microwave stimulation of peripheral nerves. Am. J. Physiol. **203**, 474, 1962.
342. *McAfee, R. D.*: Analeptic effect of microwave irradiation on experimental animals. IEEE Trans. Microwave Theory Techniques, vol. **MTT-19**, 251, 1971.
343. *McAfee, R. D.*: Microwaves stimulation of the sympathetic nervous system. Biomed. Sci. Instrum. **1**, 167, 1963.
344. *McAfee, R. D.*: Neurophysiological effects of microwave irradiation. In [518], p. 315.
345. *McAfee, R. D., Burger, C., Pizzolato, C.*: Neurological effect of 3 cm microwave irradiation. In [427], p. 251.
346. *McAfee, R. D.*: Neurophysiological effect of 3 cm microwave radiation, Am. J. Physiol. **200**, 192, 1961.
347. *McAfee, R. D.*: Physiological effects of thermode and microwave stimulation of peripheral nerves. Am. J. Physiol. **203**, 374, 1962.
348. *McCarthy, J. D.*: Radiation hazards in high-power-radars. Safety Maintenance **126**, 46 and 62, 1963.
348a. *McLaughlin, J. D.*: Tissue destruction and death from microwave radiation (radar). Calif. Med. **86**, 336, 1957. (See also *McLaughlin, J. D.*: Health hazards from microwave radiation. Western Med. **3**, 126, 1962.)
349. Merkblatt über Gesundheitsschäden durch Radargeräte und ähliche Anlagen und deren Verhütung. Deutche Gesellschaft für Ortung und Navigation. Düsseldorf, 1962.
(Note on health damage by radar and similar equipment and its prevention)
350. *Merola, L. O., Kinoshita, J. H.*: Changes on the ascorbic acid content in lenses of rabbit eyes exposed to microwave radiation. In [22], p. 285. (See also *Kinoshita, J. H., Merola, L. O., Dikmak, E., Carpenter, R. L.*: Doc. Ophthalmol. **20**, 91, 1966.)
351. *Michaelson, S. M.*: Microwave hazards evaluation — concepts and criteria. J. Microwave Power **4**, 114, 1969.
352. *Michaelson, S. M.*: The tri-service program—tribute to George M. Knauf USAF (Mc). IEEE Trans. Microwave Theory Techniques, **MTT-19**, 131, 1971.
353. *Michaelson, S. M.*: Biological effects of microwave exposure. In [82], p. 35.
354. *Michaelson, S. M.*: Human exposure to nonionizing radiant energy — potential hazards and safety standards. Proc. IEEE **60**, 389, 1972.
355. *Michaelson, S. M.*: Panel discussion. In [82], p. 256.
356. *Michaelson, S. M., Howland, J. W., Thomson, R. A. E., Mermagen, H.*: Comparison of responses to 2800 Mc and 200 Mc microwaves on increased environmental temperature. In [518], p. 162.
357. *Michaelson, S. M., Thomson, R. A. E., Quinlan, W. J.*: Effects of electromagnetic radiations on physiologic responses. Aerospace Med. **37**, 292, 1966.
358. *Michaelson, S. M., Thomson, R. A. E., Quinlan, W. J.*: Effects of electromagnetic radiations on physiologic responses. Aerospace Med. **38**, 293, 1968.

359. *Michaelson, S. M., Thomson, R. A. E., Howland, J. W.*: Thermal response in the dog exposed to microwaves. Physiologist **5**, 182, 1962.
360. *Michaelson, S. M., Thomson, R. A. E., Odland, L. T., Howland, J. W.*: The influence of microwaves on ionizing radiation exposure. Aerospace Med. **34**, 111, 1963.
361. *Michaelson, S. M., Thomson, R. A. E., Quinlan, W. J., May, B. S., Odland, L. T., Krasavage, N. J., Howland, J. W.*: The effects of microwaves on the response to ionizing radiation. Aerospace Med. **33**, 345, 1962.
362. *Michaelson, S. M., Thomson, R. A. E., Tamami, M. Y. E., Seth, H. S., Howland, J. W.*: The hematologic effects of microwave exposure. Aerospace Med. **35**, 824, 1964.
363. *Michaelson, S. M., Thomson, R. A. E., Howland, J. W.*: Physiologic aspects of microwave irradiation of animals. Am. J. Physiol. **201**, 351, 1961.
364. *Michaelson, S. M., Thomson, R. A. E., Odland, L. T., Krasawage, W., Howland, J. W.*: Tolerance of dogs to microwave exposure under various conditions. Ind. Med. Surg. **30**, 298, 1961.
365. *Mickey, G. H.*: Electromagnetism and its effects on the organism. N. Y. J. Med. **63**, 1935, 1963.
366. Microwave Radiation Hazards. California State Department of Public Health. Bureau of Occupational Health. Berkeley, Calif., 1964.
367. *Mielczarek, H.*: Neurologiczny zespół tzw. choroby mikrofalowej. Lek. Wojsk. **47**, 442, 1971. (The neurological syndrome of the so-called microwave illness)
368. *Mikołajczyk, H.*: Reakcje hormonalne i zmiany w gruczołach dokrewnych pod wpływem mikrofal. Med. Lotnicza, 39, 1972, p. 39. (Hormonal responses and changes in endocrine glands induced by microwaves)
369. *Mikołajczyk, H.*: Podziały mitotyczne komórek nabłonkowych rogówki oka u zwierząt doświadczalnych poddanych działaniu mikrofal. Med. Pracy, **21**, 15, 1970. (Mitoses of corneal epithelial cells in microwave—irradiated animals)
370. *Mikołajczyk, H.*: Dotychczasowe wyniki badań i dalsze zamierzenia w problematyce biologicznych efektów promieniowania mikrofalowego prowadzonych w Zakładzie Szkodliwości Fizycznych Instytutu Medycyny Pracy w Łodzi. Med. Lotnicza 39, 1972, p. 95. (Results of studies and further research projects on biological effects of microwaves conducted in the Department of Physical Factor Hazards of the Institute of Industrial Medicine in Łodz)
371. *Mills, W. A., Tell, R. A., Janes, D. E., Hodge, D. M.*: Nonionizing radiation in the environment. Third Ann. Natl. Conference on Radiation Control, 1972, p. 200.
372. *Mills, W. M.*: A program to study the effects of microwave radiation on various biological systems. J. Microwave Power **6**, 141, 1971.
373. *Milroy, W. C., Michaelson, S. M.*: Microwave cataractogenesis: a critical review of the literature. Aerospace Med. **43**, 67, 1972.
374. *Minecki, L.*: Objawy kliniczne u ludzi narażonych zawodowo na działanie promieniowania elektromagnetycznego wielkiej częstotliwości. Med. Pracy **16**, 300, 1965. (Clinical symptoms in personnel exposed professionally to electromagnetic very high frequency radiation)
375. *Minecki, L.*: Mutagenne działanie promieniowania elektromagnetycznego wielkiej częstotliwości. Med. Pracy **18**, 377, 1967. (Mutagenic effects of very high frequency radiation)
376. *Minecki, L.*: Ocena działania biologicznego pól elektromagnetycznych wielkiej częstotliwości jako zewnętrznego czynnika szkodliwego. Med. Pracy **14**, 75, 1963. (Evaluation of biological effects of very high frequency fields as an external hazardous influence)
377. *Minecki, L.*: Stan zdrowia ludzi narażonych na działanie pól elektromagnetycznych wielkiej częstotliwości. Med. Pracy **12**, 329, 1961. (Health status of men exposed to very high frequency electromagnetic radiation)

378. *Minecki, L.*: Promieniowanie elektromagnetyczne wielkiej częstotliwości. Działanie biologiczne i ochrona zdrowia. Wydawnictwo Związkowe CRZZ. Warsaw, 1967.
(Very high frequency electromagnetic radiation)
379. *Minecki, L., Bilski, R.*: Zmiany histopatologiczne w narządach wewnętrznych myszy poddawanych działaniu mikrofal (pasmo S). Med. Pracy **12**, 337, 1961.
(Histopathologic lesions in interal organs of mice exposed to microwave (S band))
380. *Minecki, L.*: Critical evaluation of maximum permissible levels of microwave radiation. Archiv za Higienu Rada i Toksikol. **15**, 47, 1964.
381. *Minecki, L.*: Działanie pól elektromagnetycznych wielkiej częstotliwości na rozwój embrionalny. Med. Pracy **15**, 391, 1964.
(Influence of very high frequency electromagnetic fields on embryonic development)
382. *Minecki, L., Ołubek, K., Romaniuk, A.*: Zmiany czynności odruchowo-warunkowej szczurów pod wpływem działania mikrofal (pasmo S). I. Jednorazowe działanie mikrofal. Med. Pracy **73**, 255, 1962.
(Changes in conditioned reflexes in rats exposed to microwave (S band). I. Effects of single exposures)
383. *Minecki, L., Romaniuk, A.*: Zmiany czynności odruchowo-warunkowej szczurów pod wpływem działania mikrofal (pasmo S). II. Przewlekłe działanie mikrofal. Med. Pracy **14**, 361, 1963.
(Changes in conditioned reflexes in rats exposed to microwaves (S band). II. Effects of chronic exposure)
384. Ministry of Health and Social Welfare of the Polish People's Republic: Wyjaśnienia interpretacyjne do rozporządzenia Rady Ministrów z dn. 20.10.1961.
Ministerstwo Zdrowia i Opieki Społecznej, Departament Sanitarno-Epidemiologiczny nr EP-44647-31/6629. Apr. 1966.
(Comments on interpretation of the Order of the Council of Ministers from Oct. 20, 1961)
385. *Miś, M.*: Wpływ adaptacji narządu wzroku do ciemności, odległości, pór dnia oraz sztucznie wywoływanych wad refrakcji, różnomiarowość oczu i heterotropizm w rozpoznawaniu zjawisk świetlnych ekranu radarowego. Lek. Wojsk. **37**, 7, 1961.
(The influence of visual adaptation to darkness, distance, time of the day and artificial refraction defects, unequal vision and heterotropism on recognition of light points on radar screen)
386. *Miś, M.*: Kilka uwag o reakcji narządu wzroku na zjawiska świetlne powstające na ekranie radarowym. Lek. Wojsk. **37**, 566, 1961.
(Comments on visual reaction to light phenomena on radar screen)
387. *Miro, L.*: Modifications hématoloques et troubles cliniques observés chez le personnel exposé aux ondes émises par les radars. Rev. Med. Aeron. **1**, 16, 1962.
(Hematologic changes and clinical symptoms in personnel exposed to radar emanations)
388. *Miro, L., Loubiere, R., Pfister, A.*: Recherches sur les lesions viscerales observies chez les souriés et les rats exposé aux ondes ultracourtes. Étude particuliere des effects de les ondes sur la reproduction des animeaux. Rev. Med. Aeron. **4**, 37, 1965.
(Visceral lesions observed in mice and rats exposed to ultrashort waves. Particular study of fertility)
389. *Miro, L., Loubiere, R., Pfister, A.*: Modifications morphologiques et metaboliques observees experimentalement sous l'action des champs electro-magnetiques hyperfrequence. Rev. Med. Aeron. **5**, 9, 1966.
(Morphologic and metabolic changes observed experimentally following exposure to high-frequency electromagnetic fields)
390. *Murphy, A. J., Paul, W. D., Hines, H. N.*: A comparative study of the temperature changes produced by various thermogenic agents. Arch. Phys. Med. **31**, 151, 1950.
391. *Miro, L., Pfister, A., Deltour, G., Atlan, H., Arnand, Y.*: Effect radioprotecteur des ondes

ultracourtes vis a vis des bacteries. Congress of Medical Electronics Tours. July 7, 1967. Preprint C.E.R.M.A. 5ᵇ, av. de la Porte de Sèvres, Paris, see also Rev. Med. Aeron. **4**, 21, 1965.
(Radioprotective effect of ultrashort waves in bacteria)

392. *Mittelman, J.*: Relationship between heat sensation and high frequency power absorption by humans. Digest Intern. Conf. Med. Electron. **26**, 3, 1961.

393. *Mirutenko, V. J.*: Vyvshenie miscevoi tieplovoi diji elektromagnitnych chvyl dovznynoju 3 cm na tvarym. Fiziol. Zhurnal (Kiev) **8**, 382, 1962.
(Local temperature effects of electromagnetic 3 cm waves)

394. *Mirutenko, V. J.*: Teplevoi effekt deistvija SVTch elektromagnitnogo polja na zhivotnych i niekotoryje voprosy dozimetrii SVTch-polja. In [195], p. 62 (also thesis, same title, Kiev, 1963).
(Thermal effects of exposure of animals to SHF electromagnetic fields and certain dosimetric problems)

395. *Moore, R. L., Smith, S. W., Cloke, R. L., Brown, D. G.*: A comparison of microwave detection instruments. U. S. Department of Health, Education and Welfare. Bureau of Radiological Health. Rep. BRH/DEP 70-7. Rockville, Md. 20852, 1970.

396. *Mosinger, M., Bisschop, G.*: Sur les reactions histologiques consecutives á l'irradiation par les micro-ondes de pièceo metalliques intratissulaires. C. R. Seances Soc. Biol. Filiales Associeés **154**, 1016, 1960.
(On histologic reactions caused by irradiation of intratissular metallic objects with microwaves)

397. *Mumford, W. W.*: Some technical aspects of microwave radiation hazards. Proc. IRE **49**, 427, 1961.

398. *Mumford, W. W.*: Heat stress due to RF radiation. Proc. IEEE **57**, 171, 1969.

399. *Mumford, W. W.*: Heat stress due to RF radiation. In [82], p. 21.

400. *Mumford, W. W.*: Heat stress due to RF radiation. Non-ionizing radiation **1**, 113, 1969.

401. *Murray, R., Abraham, J. D. R., Chambers, J. H., Elliot, P. M., French, G. E. Gilbert, R. R., Holden, H., Muirhead, A.*: How safe are microwaves? Non-ionizing Radiation **1**, 7, 1969.

402. News Editorial: Microwave radiation called growing hazard. Electron. Design. **14**, 28, 1969.

403. *Niepołomski, V., Smigla, K.*: Patomorfologia narządów zwierząt doświadczalnych poddanych działaniu pola elektromagnetycznego o częstotliwości 10.7 MHz. Abstr. Symp. Med. Przem. Katowice, 27–29, November 1964.
(Pathomorphology of experimental animals exposed to 10.7-MHz electromagnetic field)

404. *Nikogosjan, S. V.*: Vlijanie SVTch na aktivnost cholinesterazy v syvorotkie krovi i organach zivotnych. In: O biologitcheskom vozdiejstvii svierchvysokich tchastot. Moscow, 1960, p. 81.
(The influence of SHF on blood serum and organ cholinesterase activity)

405. *Nikogosjan, S. V.*: Vlijanie santimetrovych i decimetrovych voln na soderzhanie belka i belkovych frakcji v syvorotko krovi zhivotnych. In: Voprosy biologitcheskogo dieistvija sverchvysokotchastotnogo (SVTch) elektromagnitnogo polja. Tez. Nautch. Konf. Leningrad, 1962, p. 33.
(The influence of centimeter and decimeter waves on blood serum proteva content and fractions in animals)

406. *Nikonova, K. V.*: Vlijanie elektromagnitnogo polja wysokoi tchastoty na krovjanoje davlenie i temperaturu tela eksperimentalnych zhivotnych. In: O biologitcheskom deistvii elektrom. polei. radiotchastot. Medgiz. Moscow, 1964, p. 49.
(The influence of electromagnetic high-frequency fields on blood pressure and body temperature in animals)

407. *Oick, R., Horai, H.*: Fatal dosage of microwaves to the mouse and body temperature increase during exposure. Eoei (Natl. Defense J., Tokyo), **9**, 75, 1962 (English Summary).

408. *Okress, E. C.* (ed.) Microwave power engineering, vol. I and II. Academic Press, New York, 1968.
409. *Oldendorf, W. H.*: Focal neurological lesions produced by microwave irradiation. Proc. Soc. Exp. Biol. Med. **72**, 432, 1949.
409a. *Ornowski, M.*: Badania nad wpływem mikrofal na narządy wewnętrzne. Med. Lotnicza **20**, 47, 1967.
 (Investigations on the influence of microwaves on internal organs)
410. *Osborne, C.*: Studies on the biological effects of 200 Mc. In [418], p. 20.
411. *Osborn, S. L., Frederick, J. N.*: Microwave radiations heating of human and animal tissues by means of high-frequency current with wavelength of twelve centimeters. J. Am. Med. Assoc. **137**, 1031, 1948.
412. *Osipow, J, A., Kulikovskaja, E. L., Kaljada, T. V.*: Uslovija oblutchenija elektromagnitnym polem SVTch rabotajushtchich na nastrojke i ispytanii radiotechnischeskich priborov. Gig. Sanitaria **2**, 100, 1962.
 (Irradiation by electromagnetic fields of persons engaged in tuning of radiotechnical instruments)
413. *Ostachowicz, M.*: Wyniki badania oczu królików napromieniowanych falami radarowymi. Lek. Wojsk. **39**, 9, 1963.
 (Results of examination of eyes of rabbits irradiated with radar waves)
414. *Paff, G. H., Boucek, R. J., Nieman, R. E., Deichmann, W. B.*: The embryonic heart subjected to radar. Anat. Rec. **147**, 379, 1963.
415. *Paff, G. H., Deichman, W. B., Boucek, R. J.*: The effects of microwave irradiation on the embryonic chick heart as revealed by electrocardiographic studies. Anat. Rec. **142**, 264, 1962.
416. *Palmisano, W. A., Peczenik, A.*: Some considerations of microwave hazards exposure criteria. Military Med. **131**, 611, 1966.
417. *Panecki, M.*: Podstawowe zagadnienia z techniki mikrofalowej. In [574], p. 9.
 (Basic problems of microwave techniques)
418. *Pattishal, E. G., Banghart, F. W.* (eds.): Proceedings. Second Tri-Service conference on biological effects of microwave energy. Univ. Virginia, Charlottesville Va, July 8–10, 1958. ASTIA doc. AD 131477.
419. *Pavlov, I. P.*: Conditioned reflexes. Oxford University Press Inc., New York, 1927.
420. *Pavlov, I. P.*: Croonian lecture. Proc. Royal. Soc. **103B**, 97, 1928.
421. *Pavlov, I. P.*: Lectures on conditioned reflexes. Laurence, London, 1929.
422. Performance standard for microwave ovens. Federal Register 35, No 194. Title 42-Public Health (Oct. 6, 1970).
423. *Petrov, I. P.*: Kombinirovannye vlijania na organizm SVTch elektromagnitnych izlutchemii i drugich faktorov. In [425], p. 63
 (Combined effects of SHF electromagnetic and other factors on the organism).
424. *Petrov, I. R.* (ed.): Mediko-biologitcheskije problemy SVTch-izlutchenii. Medgiz. Leningrad, 1966.
 (Medico-biological problems of SHF radiations)
425. *Petrov, I. R.* (ed.): Vlijanie SVTch-izlutchenii na organizm tcheloveka i zhivotnych. Medicina. Lenigrad, 1970.
 (English translation: Influence of microwave radiation in the organism of man and animals, *I. R. Petrov* (ed.) NASA TT-F-708, Feb. 1972. National Technical Information Service, Springfield Va.
426. *Petrov, I. R., Subbota, A. G.*: Zakljutchenie. In [425], p. 201.
 (Concluding remarks)
427. *Peyton, M. F.*, (ed.): Proceedings fourth annual tri-service conference on biological effects

of microwave radiating equipments: biological effects of microwave radiations. Plenum Publishing Corporation. New York, 1961.
428. *Pinakatt, T., Cooper, T., Richardson, A. W.*: Effect of ouabain on the circulatory response to microwave hyperthermia in rat. Aerospace Med. **34**, 497, 1963.
429. *Pinakatt, T., Richardson, A. W., Cooper, T.*: The effect of digitoxin on the circulatory response of rats to microwave irradiation. Arch. Int. Pharmacodynamie Therapie **156**, 151, 1965.
430. *Pitenin, J. W., Subbota, A. G.*: Ob obrazovanii jazvy zeludka u krolikov pri vozdejstvii SVTCh-izlutchenij na prigastralnuju oblast zivota. Bjul. Eksper. Biol. Med. **9**, 55, 1965.
(On induction of stomach ulcers in rabbits by SHF irradiation of the perigastric ventral region in rabbits)
431. *Pitenin, I. V.*: Patologonatomitcheskije izmienienia v organach i tkaniach zivotnych pri vozdejstvii elektromagnitnovo polia sverchvysokich tchastot. Tez. Nautch. Konf. Leningrad, 1962, p. 36.
(Pathological lesions in organs and tissues of animals exposed to super high-frequency electromagnetic fields)
432. *Pivovazov, M. A.*: Vlijanie SVTCh polja malych intensivnosti na neketorye analizatory tcheloveka. Abstr. Tez. Nautch. Konf. Leningrad, 1966.
(The influence of low-intensity SHF fields on certain senses in man)
433. *Pol, W.*: Zagadnienie wpływu mikrofal emitowanych z nadajników radarowych na powstawanie zaćmy. Lek. Wojsk. **38**, 318, 1962.
(The problem of the influence of microwaves emitted from radar equipment on cataractogenesis)
434. *Powell, C. H., Rose, V. E.*: Health surveillance of microwave hazards. Am. Ind. Hyg. Assoc. J. **31**, 358, 1970.
435. *Povzhitkov, V. A., Tjagin, N. V., Grebieshetchnikova, A. M.*: Vlijanie sverchvysokotchastotnogo impulsnogo elektromagnitnogo polja na zatchatie i tetchenie bieremennosti u belych myshei. Bjul. Eksper. Biol. Med. **5**, 103, 1961.
(The influence of pulsed super high-frequency fields on conception and the course of gestation in white mice)
436. *Prausnitz, S., Süsskind, C., Vogelhut, P. O.*: Longevity and cellular studies with microwaves. In [427], p. 135.
437. *Prausnitz, S., Süsskind, C, Vogelhut, P. O.*: Effects of chronic microwave irradiation on mice. IRE Trans. Bio-med. Electron. **9**, 104, 1962.
438. *Presman, A. S.*: Elektromagnitnyje polja i zhivaja priroda. Nauka. Moscow, 1968.
(English translation: Electromagnetic fields and life. Plenum Press Publishing Corporation New York, 1970.
Polish translation: Pola elektromagnetyczne, a żywa przyroda. PWN. Warsaw, 1971)
439. *Presman, A. S.*: Eksperimentalnaja ustanovka dlja doziruemogo oblutchema krolikov mikrovolnami 10-santimetrovogo diapozona. Novosti Medicinskoi Techniki **4**, 51, 1960.
(Experimental arrangement for dosed (quantified) irradiation of rabbit with 10-cm microwaves)
440. *Presman, A. S.*: Voprosy mechanisma biologitcheskogo deistvija santimetrovyvh voln. Usp. Sov. Biol. **56**, 161, 1963.
(Problems of the mechanism of biological effects of centimeter waves).
441. *Presman, A. S.*: Elektromagnitnoje polje kak gigienitcheski faktor. Gig. Sanitaria **21**, 32, 1956.
(Electromagnetic field as a hygienic factor)
442. *Presman, A. S.*: Deistvie mikrovoln na zhivye organizmy i biologitcheskije struktury. UFN **86**, 263, 1965.
(The action of microwaves on living organisms and biological structures)

443. *Presman, A. S.*: O fizitcheskich osnovach biologitcheskogo deistvija mikrovoln. Usp. Sovr. Biol. **41**, 40, 1956.
(Physical bases of biological effects of microwaves.)
444. *Presman, A. S.*, *Kamenskij, J. I.*, *Levitina, N. A.*: Biologitcheskoje deistvie mikrovoln. Usp. Sorr. Biol. **51**, 84, 1961.
(Biological effects of microwaves)
445. *Presman, A. S.*, *Levitina, N. A.* Neteplovoe deistvie mikrovoln na ritm serdetchnych sokrashtchenii u zhivotnych. I. Issledovanije deistvija nepreryvnych mikrovoln. Bjul. Eksper. Biol. Med. **53**, (1), 41, 1962.
(Nonthermal effects of microwaves on cardiac rhythm in animals. I. Influence of continuous exposure to microwaves)
446. *Presman, A. S.*, *Levitina, N. A.*: Neteplovoe deistvie mikrovoln na ritm serdetchnych sokrastchenii u zhivotnych. II. Issledovanije deistvija impulsnych mikrovoln. Bjul. Eksper. Biol. Med. **53** (2), 39, 1962.
(Nonthermal effects of microwaves on cardiac rhythm in animals. II. Influence of pulsed microwaves).
447. Radiation hazards from electronic equipment. Canadian Standards Association (CSA Standard Z 65-1965)
448. *Rae, J. W.*, *Herrick, J. F.*, *Wakim, K. G.*: A comparative study of the temperatures produced by microwaves and short-wave diathermy. Arch. Phys. Med. **30**, 199, 1949.
449. *Rheinberg, G. L.*, *Moghissi, A. A.*, *Pepper, E. W.*: Effects of microwaves on optical activity. In [82], p. 101.
450. *Richardson, A. W.*: Biologic effects of non-ionizing electromagnetic radiations. Scientia **103**, 447, 1968.
451. *Richardson, A. W.*: Effects of microwave-induced heating on blood flow through peripheral skeletal muscles. Am. J. Phys. Med. **33**, 103, 1959.
452. *Richardson, A. W.*: Review of the work conducted at the St. Louis University School of Medicine. In [418], p. 169.
453. *Richardson, A. W.*: Blood coagulation changes due to electromagnetic microwave irradiation. Blood **14**, 1237, 1959.
454. *Richardson, A. W.*, *Duane, T. D.*, *Hines, H. M.*: Experimantal cataracts produced by 3-centimeter pulsed microwave irradiation. AMA Arch. Ophthalmol. **45**, 382, 1951.
455. *Richardson, A. W.*, *Duane, T. D.*, *Hines, H. M.*: Experimental lenticular opacities produced by microwave irradiation. Arch. Phys. Med. **29**, 765, 1948.
456. *Richardson, A. W.*, *Imig, Ch. J.*, *Feucht, B. L.*: The relationship between deep tissued temperature and blood flow during electromagnetic irradiation. Arch. Phys. Med. **31**, 19, 1950.
457. *Richardson, A. W.*, *Lomax, D. H.*, *Nicholas, J.*, *Green, H. D.*: The role of energy, pupillary diameter and alloxan diabetes in the production of ocular damage by microwave irradiations. Am. J. Ophthalmol. **35**, 993, 1952.
458. *Roberts, J. E.*, *Cook, H. F.*: Microwaves on medical and biological research. Brit. J. Appl. Phys. **3**, 33, 1952.
459. *La Roche, L. P.*, *Braun, A. F.*, *Zaret, M. M.*: An operational safety program for ophthalamic hazards of microwave. Arch. Environ. Health **20**, 350, 1970.
460. Rozporządzenie Rady Ministrów z dnia 20.10.1961 r. w sprawie bezpieczeństwa i higieny pracy przy używaniu urządzeń mikrofalowych. Dziennik Ustaw Polskiej Rzeczpospolitej Ludowej Nr 48, po. 255, 1961.
(Order of the Council of Ministers of Oct. 20, 1961, concerning safety and hygiene at work when using microwave equipment)
461. Rozporządzenie Rady Ministrów z dnia 25.05.1972 r. w sprawie bezpieczeństwa i higieny

pracy przy stosowaniu urządzeń wytwarzających pola elektromagnetyczne w zakresie mikrofalowych. Dziennik Ustaw Polskiej Rzeczpospolitej Ludowej Nr 21 (8.II.1972) poz. 153.
(Order of Council of Ministers of May 25, 1972, concerning safety and hygiene when using equipment generating electromagnetic fields in the microwave region)

462. *Rubner, A. M.*: Biologic effects of radio and microwave — present knowledge: future directions. (IEEE ICC Conference Publications Volume 5, 1969, p. 32-1 to 32-6).
463. *Sacchitelli, G., Sacchitelli, E.*: Sul comportaments delle glutationemia in seguito ad irradiationi con microonde radar. Folia Medica (Naples) **41**, 342, 1958.
(On the behavior of blood glutathione in cats exposed to radar microwaves)
464. *Sacchitelli, G., Sacchitelli, E.*: L'azione delle micronde radar sulla plasmalipisi e sull'amidosi serica. Folia Medica (Naples) **39**, 1037, 1956.
(The action of radar microwaves on plasma lipose and serum amidase)
465. Safety precautions for shore activities. NAVSO P-2455. Apr. 1965 (citation according to [491]).
466. Safety regulations for working with microwave radiation. BVX 13-6. Labour Protection Department N. Y. Philips Gloeil lampenfabrieken. Eindhoven, Holland, 1967.
467. Safety precautions relating to intense radio-frequency radiation. The British Post Office. 25822 Wt. 2006-37290 K 24 4/60. Her Majesty's Stationary Office. London, 1960.
468. Safety level electromagnetic radiation with respect to personnel C-95.1. (Rep. USASI-C 95.1. United States of America Standards Institute. New York, 1966.
469. *Salati, O. M., Anne, A., Schwan, H. P.*: Radiofrequency radiation hazards. Electron. Ind. **21**, 96, 1962.
470. *Samaras, G. M., Muroff, L. R., Anderson, G. E.*: Prolongation of life during high intensity microwave exposures. IEEE Trans. on Microwave Theory Techniques, vol. **MTT-19**, 245, 1971.
471. *Sawicki, W., Ostrowski, K.*: Non-thermal effects of microwave radiation *in vitro* on peritoneal mast cells of the rat. Am J. Phys. Med. **17**, 225, 1968.
(Shortwave therapy)
472. *Schliephake, E.*: Kurzwellen-therapie. G. Thieme. Stuttgart, 1952.
(Shortwave therapy)
473. *Schwan, H. P.*: Interaction of microwave and radiofrequency radiation with biological systems. IEEE Trans. Microwave Theory Techniques, vol. **MTT-19**, 146, 1971.
474. *Schwan, H. P.*: Interaction of microwave and RF radiation with biological systems. In [82], p. 13.
475. *Schwan, H. P.*: Electrical properties of tissue and cell suspension. Adv. Biol. Med. Phys. **5**, 147, 1957.
476. *Schwan, H. P.*: SVTch-biofizika. In E. Okress (ed.) "SVTch-Energetika" vol. 3 chapter 5.2, p. 7. Izd. "Mir" Moskva, 1971.
(Russian translation of "Microwave power engineering" E. C. Okress (ed.) vol. 2, Applications, Academic Press, Inc. New York, 1968.
477. *Schwan, H. P.*: Biophysics of diathermy. In: Therapeutic Heat and Cold. Licht. (ed.) New Haven, 1965, sec. 12, p. 310.
478. *Schwan, H. P., Anne, A., Saito, M., Salati, O. M.*: Relative microwave absorption cross sections of biological significance. In [427], p. 153.
479. *Schwan, H. P., Li, K.*: Variations between measured and biologically effective microwave diathermy dosage. Arch. Phys. Med. Rehab. **36**, 363, 1955.
480. *Schwan, H. P., Li, K.*: Hazards due to total body irradiation by radar. Proc. IRE **44**, 1572, 1965.
481. *Schwan, H. P., Li, K.*: The mechanism of absorption of ultrahigh frequency electromagnetic energy in tissue, as related to the problem of tolerance dosage. IRE Trans Med. Electron., vol. **PGME-4**, 45, 1956.

482. *Schwan, H. P., Li, K.*: Capacity and conductivity of body tissues at ultrahigh frequencies. Proc. IRE **41**, 1735, 1953.
483. *Schwan, H. P., Piersol, G. M.*: The absorption of electromagnetic energy in body tissues, a review and critical analysis. Part I. Biophysical aspects. Am. J. Phys. Med. **33**, 371, 1954.
484. *Schwan, H. P., Piersol, G. M.*: The absorption of electromagnetic energy in body tissues, a review and critical analysis. Part II. Physiological and clinical aspects. Am. J. Phys. Med. **34**, 425, 1955.
485. *Searle, G. W., Imig, C. J., Dahlen, R. W*: Studies with 2450 Mc CW exposure to the heads of dogs. In [418], p. 54.
486. *Searle, G. W., Dahlen, R., Imig, G., Mundee, C., Thomson, J., Thomas, J., Moressi, W.*: Effects of 2450 microwaves in dogs, rats and larvae of the common fruit fly. In [427], p. 187.
487. *de Seguin, L.*: Lois de la repartition de la chaleur dans les tissus organiques apres irradiation per un champ de micro-ondes. C. R. Acad. Sci. Phys. Biol. **228**, 135, 1949.
(The law of heat distribution in organic tissues following irradiation with microwave fields)
488. *de Seguin, L.*: Reversibilité des lesions observées sur les petits animeaux exposé a des ondes d'ultracourte frequence (longuer d'onde 21 cm). C. R. Acad. Sci. **225**, 76, 1949.
(Reversibility of lesions observed in small animals exposed to ultra high-frequency waves 21-cm wavelength)
489. *Semonov, A. I.*: O vlijanii SVTch elektromagnitnogo polia na temperaturu tkanei biedra krolika. Thesis. Leningrad, 1965.
(On the influence of SHF electromagnetic field on the temperature reaction of hip tissues in the rabbit)
490. *Seth, H. S., Michaelson, S. M.*: Microwave hazards evaluation. Aerospace Med. **35**, 734, 1964.
491. *Setler, L. R., Snavely, D. R., Solem, D. L., Van Wye, R. F.*: Regulations, standards and guides for microwaves, ultraviolet radiation and radiation from lasers and television receivers an annotated bibliography. U. S. Department Health, Education, and Welfare. Public Health Service. Consumer Protection and Environmental Health Service. Environmental Control Administration. Bureau of Radiological Health. Rockville, Md 20852. (Public Health Service Publication No 999-RH-35) April 1969. Section V. Microwaves, pp. 45-79.
492. *Sevastianov, V. V.*: Ob izmenenijach intensivnosti SVTch elektromagnitnych izlutchenii v sviazi s zadetchej ih gegenitcheskoj ocenki. Voj. Med. Zhurn. **7**, 21, 1965.
(On measurements of SHF electromagnetic radiation intensity related to hygienic evaluation)
493. *Shapiro, A. R., Lutomirski, R. F., Yura, H. T.*: Induced fields and heating within a cranial structure irradiated by an electromagnetic plane wave. IEEE Trans. Microwave Theory Techniques, vol. **MTT-19**, 187, 1971.
494. *Sluce, P. M.*: Simple methods for determining energy distribution in a microwave oven. Nonionizing Radiation **1**, 131, 1969.
495. *Siekierzyński, M.*: Wpływ promieniowania mikrofalowego na metabolizm żelaza u królików. Med. Lotnicza 39, 1972, p. 53.
(The influence of microwave radiation on iron metabolism in rabbits).
496. *Siekierzyński, M.*: Wpływ promieniowania mikrofalowego na czynność erytropoetyczną i metabolizm żelaza u królików. Doctor's thesis. Instytut Kształcenia Podyplomowego WAM. Warszawa, 1970.
(The influence of microwave radiation on erythropoiesis and iron metabolism in rabbits)
497. *Sigler, A. T., Lillienfeld, A. M., Cohen, B. H., Westlake, J. E.*: Johns Hopkins Hosp. Bull. **117**, 374, 1965.

498. *Singatulina, R. G.*: Vlijanie tokov ultravysokoi tchastoty na bielkovyje frakcji syvorotki krovi. Bjul. Eksper. Biol. Med. **7**, 64, 1961.
 (The influence of ultra high-frequency currents on blood protein fractions)
499. *Slabospitalskii, A. A.*: K voprosu o mikrovolnych porazhenijach kozhi. In [195], p. 92
500. *Slabospitalskii, A. A.*: Do pytania po mechanizmu vplyvu mikrochvyl na shkiru. Fiziol. Zhurnal **11**, 225, 1965 (Kiev).
 (On the problem of the mechanism of influence of microwaves on the skin)
501. *Smirnova, M. I., Sadtchikova, M. N.*: Opredelenie funkcionalnoi aktivnosti shtchytovidnoi zhelezu s pomoshtchju radioaktivnogo joda u rabotajushtchych z generatorami SVTch. Trudy Inst. Gig. Truda i Profzabolevanii AMN, SSSR, 1960, p. 50.
 (Determination of functional activity of the thyroid gland radioiodine in persons working with SHF equipment)
502. *Sokoloff, L.*: Metabolism of the central nervous system *in vivo*. In Handbook of physiology, sec. 1. vol. 3, Washington, American Physiological Society. D. C., 1960, p. 1843.
503. *Solem, D. L., Remak, D. G., Moore, R. L., Crawford, E. E., Rechen, H. I. L.*: Preliminary measurements of electromagnetic radiation fields near microwave ovens. Non-ionizing Radiation **1**, 88, 1969.
504. *Spalding, J. F., Freyman, R. W., Holland, L. M.*: Effects of 800 MHz electromagnetic radiation on body weight, activity of hematopoiesis and life span in mice. Health Phys. **20**, 421, 1971.
505. *Spector, W. S.* (ed.): Handbook of biological data. WADC Technical Report 56 273 (AD 110501). Wright-Patterson Air Force Base. Ohio, 1958.
506. *Stodolnik-Barańska, W.*: Wpływ wibracji i mikrofal na zachowanie się komórek i chromosomów. Doctors Thesis. Akademia Medyczna, Warsaw. 1966
 (The influence of vibration and microwaves on cells and chromosomes)
507. *Stodolnik-Barańska, W.*: Microwave induced lymphoblastoid transformation of human lymphocytes *in vitro*. Nature **214**, 202, 1967.
507. a. *Stratton, J. A.*: Electromagnetic Theory, Mc Graw-Hill Book Company. New York, 1941, p. 428.
508. *Strollo, M.*: Un tipo di analisi di lavoro per il personel della mamutenzione dei ponti-radio nel quadro della medicina preventiva del lavoro. Riv. Med. Aeronat. Spaziale **30**, 41, 1967.
 (Work analysis of radio-tower maintenance staff in occupational preventive medicine)
509. *Subbota, A. G.*: Sistema organov pishtchevarneija. In [425], p. 47.
 (Digestive tract)
510. *Subbota, A. G.*: Nietieplavoje diestvie mikroradiovoln na organizm. Voenno. Med. Zhurnal **9**, 39, 1970.
 (Nonthermal effects of microwaves in the organism)
511. *Subbota, A. G.*: Sistema vnieshnego dychania, serdetchnososudistaja sistema i krov. In [425], p. 44.
 (Respiratory system, cardiovascular system and blood)
512. *Subbota, A. G.*: O vlijanii impulsnogo SVTch-elektromagnitnogo polja na vyshashuju nervnuju dejatelnost sobak. Bjul. Eksper. Biol. Med. **10**, 55, 1958.
 (On the influence of pulsed SHF electromagnetic fields on higher nervous function in dogs)
513. *Subbota, A. G.*: O vlijanii SVTch-elektromagnitnogo polja na vyzhshuju nervnuju dejatelnost sobak. Trudy VMA im. S.M. Kirova. **73**, 35, 1957, see also pp. 78 and 127.
 (On the influence of SHF electromagnetic fields on the higher nervous function in dogs)
514. *Subbota, A. G.*: Izmenenie funkcji razlitchnych sistem organizma. In [425], p. 70.
 (Functional disturbances in various systems of the organism)
515. *Subbota, A. G.*: In [424], p. 38 (quoted according to [425]).
516. *Subbota, A. G.*: Tezisy dokladov XVI Ukrainskoi respubl. nautchnotechnitcheskoi konferencii posvjashtchennoi Dnju radio. Kiev, 1966, p. 148 (cited according to [425]).

517. *Subbota, A. G., Svetlova, Z. P.*: O dezataptizujushtchem i dekompensizujushtchem deistvii mikrovoln. Abstr. Sym. Gig. Truda i Biologitcheskogo Deistvie Elektromagnitnych Voln Radiotchastot, Moscow, 1972, p. 13.
(On desadaptation and decompensation induced by microwaves)
518. *Süsskind, C.* (ed.): Proceedings Third Annual Tri-Service Conference on Biological Hazards of Microwave Radiating Equipments. Univ. of California. Berkeley, Calif. 25—27. Aug. 1959.
519. *Süsskind, C., Vogelhut, P. Q.*: Cavity perturbation measurement of the effects of microwave radiation on proteins. IRE Trans. Med. Electron. **5**, 668, 1962.
520. *Swanson, J. R., Rose, V, E., Powell, C. H.*: A review of international microwave quides. Am. J. Ind. Hyg. Assoc. J. **31**, 623, 1970.
521. *Święcicki, W.*: Zachowanie się białek w surowicy krwi świnek morskich narażonych na termiczne działanie mikrofal. Med. Pracy, **18**, 471, 1967.
(Serum blood proteins in guinea pigs exposed to thermal microwave doses)
522. *Święcicki, W., Szmigielski, S.*: Wpływ mikrofal na poziom glutationu. Lek. Wojsk. **48**, 17, 1972.
(Influence of microwaves on glutathione levels)
523. *Święcicki, W., Edelwejn, Z.*: Obraz elektroforetyczny białek osocza krwi u królików poddanych ostremu napromieniowywaniu falami elektromagnetycznymi bardzo wielkiej częstotliwości. Farmacja Polska **19**, 189, 1963.
(Electrophoresis of blood protein in rabbits exposed to acute irradiation with very high frequency electromagnetic waves)
524. *Święcicki, W., Edelwejn, Z.*; Wpływ napromieniowania mikrofalami w paśmie trzech i dziesięciu centymetrów na obraz białek krwi u królików. Med. Lotnicza **11**, 54, 1963.
(The influence of 3-cm and 10-cm microwave irradiation on blood proteins in rabbits)
525. *Svetlova, Z. P.*: Trudy VMA im. S. M. Kirova, **166**, 38, 1966 (cited according to [425]).
526. *Svetlova, Z. P.*: In: Gigiena Truda i biologitcheskoje deistvie elektromagnitnych polej radiotchastot. Annot. dokl. Moscow, 1963, p. 81 (cited according to [425]).
527. *Svetlova, Z. P.*: Ob izmienieniach simetritchnych uslovnych i bezuslovnych refleksov u sobak pri vozdejstvii SVTch polja decimetrovogo diapozona. In: Voprosy biologitcheskogo deistvija sverchwysoko-tchastnotnogo (SVTch) elektromagnitnogo polja. Tezisy. Leningrad, 1962, p. 43.
(On changes in symmetrical unconditioned and conditional reflexes in dogs exposed to decimeter SHF fields)
528. *Sviridov, L. P.*: Vlijanie sverchvysotnych izlutchenii razlitchenoi tchastoty na tetchnie salmonelloznoi infekcji u belych myshei. Zh. Microbiol. Epidemiol. Immunobiol. **47**, 10, 1970.
(The effect of ultra-high-frequency radiation of various intensities on the course of *Salmonella* infections in white mice)
529. *Syngajevskaja, V. A.*: Izmenenija obmena vetchestv. In [425], p. 51.
(Metabolic changes)
530. *Syngajevskaja, V. A.*: Lekcii po probleme SVTch-izlutchenii. VMA. Leningrad, 1964.
(Lectures on SHF radiation problems)
531. *Syngajevskaja, V. A.*: In: Gigiena truda i biologitcheskoje deistvie elektromagnitnych voln radiotchastot. Moscow, 1968, p. 150
532. *Syngajevskaja, V. A., Ignatieva, O. S., Pliskina, T. P.*: In: Gigiena truda i biologitcheskoje deistvie elektromagnitnych polei radiotchastot. Annot. dokladov (notes from conference) Moscow, 1963, p. 96 (cited according to [529]).
533. *Syngajevskaja, V. A., Ignatieva, O. S., Pliskina, T. P.*: Vlijanie SVTch-oblutchenija metrovogo i decimetrovogo diapozonov na endokrinnuju regulaciju uglevodnovogo obmena i funkcjonalnojesostojanie nadpotchetchnikov u krolikov i sobak. In: Voprosy biologitcheskogo deistvija sverchvysokotchastotnogo (SVTch) elektromagnitnogo polja. Tez. Nautch. Konf. Leningrad, 1962, p. 52.

(On the influence of SHF irradiation in the meter and decimeter range, on the endocrine regulation carbohydrate metabolism and functional state of adrenals in rabbits and dogs).
534. *Shimkovitch, I. S., Shilajev, V. G.*: Katarakta oboich glaz rozvivajuschtchoja v rezultacie kratkovremennogo prebyvanii v wverchwysokotchastotnym elektromagnitnym polje wysokoi platnosti. Vestnik Oftalmologii **4**, 12, 1959.
(Bilateral cataract induced by a short time exposure to high-intensity super high-frequency electromagnetic field)
535. *Szmigielski, S.*: Wpływ przewlekłego napromieniowania mikrofalowego na granulopoezę, Med. Lotnicza **2**, p. 89, 1968.
(The influence of chronic microwave irradiation on granulopoiesis)
536. *Tallarics, R., Ketchum, J.*: Effect of microwaves on certain behavior patterns of the rat. In [518], p. 74.
537. *Tanner, J. A., Romero-Sierra, C.*: Bird feathers as sensory detectors of microwave fields. In [82], p. 185.
538. *Taylor, L. S.*: Radiation protection trends in the United States. Health Phys. **20**, 499, 1971.
539. *Taylor, L. S.*: The development of exposure guidelines. Proc. Conf. on Estimation of Low-level Radiation Effects in Human Population (Argonne Nat. Lab. Dec. 1970. Rep. ANL-7811). May 1971, p. 27.
540. *Techepikova, I. P.*: In Materialy vsesojuznogo zjezda fizjoter. i kurortologii. Baku, 1965, p. 158 (cited according to [511]).
541. *Tchepikova, I. P.*: In: Gigiena truda i biologitcheskoe deistvie elektromagnitnych polei radiotchastot. Annot. dokladov (notes from lectures). Moscow, 1963 (cited according to [511]).
542. *Tchuchlovin, B. A.*: Izmenenija immunologitcheskoi reaktivnosti organizma, svistv bakterii, virusov i prosteishych. In [425], p. 93.
(Changes in immunologic reactivity of the organism, properties of bacteria, viruses, and protozoa)
543. Technical manual for radiofrequency radiation hazards. NAVSHIPS 0900-005-8000. July, 1966 (citation according to [491]).
544. *Teixeiria-Pinto, A. A., Nejelsky, L. L., Cutler, J. L., Heller, J. H.*; The behaviour of unicellular organisms in an electromagnetic field. Exp. Cell. Res. **20**, 548, 1960.
545. *Tell, R. A., Kinn, J. B.*: Ocular heating during dental diathermy. J. Oral Surg. **30**, 284, 1972.
546. *Tell, R. A.*: Microwave energy absorption in tissue. Environmental Protection Agency, Office of Research and Monitoring, Washington, D. C., 1972 (Febr.) (Twinbrook Research Labo-Ratory technical publication), see also: Reference data for radiofreguency emission hazard analysis. U. S. Environmental Protection Agency. Office of Radiation Programs. Washington, D. C., June 1972, ORP/SID 72-3.
547. *Tengroth, B., Aurell, E.*: Retinal lesions as the result of microwave exposure. IIIrd Int. Conference Medical Physics. Göteborg, 1972, Abstr. 32.2
548. *Tettoni, E., Comino, E.*: Intorno agli effetti delle onde decimetriche (radarterapia) sulla motilita gastroenterica delle cavie. Minerve Fisioter. Radiol. **5**, 135, 1960.
(On the subject of the effects of decimeter waves [radar therapy] on gastroenteric mobility in guinea pigs).
549. *Thomson, R. A. E., Michaelson, S. M., Howland, J. W.*: Microwave radiation and its effect on response to X radiation. Aerospace Med. **37**, 304, 1966.
550. *Tibbs, C. E.*: Radiation safeguards for microwave ovens. Non-ionizing Radiation **1**, 73, 1969.
551. *Tjagin, N. V.*: Niekotoryje voprosy medicinskogo obespietchenia litchnogo sostava radiotechnitcheskoi sluzby. Voj. Med. Zhurnal **9**, 14, 1960.
(Certain problems of medical surveillance of the personnel of the radiotechnical service)
552. *Tjagin, N. V.*: Trudy VMA 73, 102, 1957 (quoted according to [425])

553. *Tjagin, N. V.*: O teplovom deistvii SVTch-elektromagnitnogo polja. Bjul. Eksper. Biol. Med. **64**, 67, 1958.
(On thermal effects of SHF electromagnetic fields)
554. *Tolgskaja, M. C., Gordon, Z. W.*: Morfologitcheskoje izmenenija pri deistvii elektromagnitnych voln radiotchastot (eksperimentalnye issledovania). Izd. Medicina. Moscow, 1971.
(Morphologic changes induced by electromagnetic radiofrequency waves. Experimental investigations)
555. *Tolgskaja, S. M., Gordon, Z. W., Markov, V. V., Voroncov, R. S.*: Vlijanije chronitcheskogo mikrovolnogo oblutchenia v intermittizujushtchem i neprerywnom rezhime na neizosekreternuju funkcju gipotalamusa. In: Gigiena truda i biologitcheskoje deistvie elektromagnitnych voln radiotchastot. Sym. Abstr. Moscow, 1972, p. 34.
(The influence of chronic intermittent and continuous microwave irradiation on the hypothalamic neurosecretory function)
556. *Tolgskaja, M. S., Gordon, Z. W., Lobanova, E. A.*: Morfologitcheskoje izmienienija u eksperimentalnych zhivotnych pri vozdeistvii impulsnych i niepreryvnych SVTch. In [294], p. 90.
(Morphologic changes in experimental animals exposed to pulsed and continuous SHF)
557. *Tolgskaja, M. S., Nikonova, K. V.*: Gistologitcheskie izmienienia w organach belych kryz pri chrontcheskom vozdeistvii elektromagnitnych vozdeistvii. In O biologitcheskom deistvii elektromagnitnych polei radiotchastot. Moscow, 1964, p. 89.
(Histological changes in organs of white rats following chronic exposure to electromagnetic fields of super high-frequency)
558. *Tomberg, V.*: Specific thermal effects of high-frequency fields. In [427], p. 221.
559. Threshold limit values of physical agents with intended changes adopted by ACGIH for 1971. (American Conference of Governmental and Industrial Hygienists) Publ. Cincinnati, Ohio, 1971.
560. *Trojanskij, M. P., Kuglikov, T. J. Kornikov, R. M., Kolashnikov, L.*: Nieketoryje itogi izutchenija sostojania zdorovia specjalistov, rabotajushtchich s generatorami SVTch. Voj. Med. Zhurnal **7**, 30, 1967.
(Certain problem of investigations on the health status of specialists working with microwave sources)
561. U. S. Air Force: Health information of microwave radiation. APF 160-6-18, USAF Surgeon General. Washington, D.C. 1958.
562. *Uspenskaja, N. V., Rozencvit, G. E., Kudriavceva, S. V.*: Klinika chronitcheskogo vozdeistvia elektromagnitnych voln sverch vysokoi tchastoty. Informacjonnoe Pismo. Leningrad, 1960. Min. Zdrovochranenija RSFSR.
(Studies of chronic exposure to electromagnetic waves of super high-frequency. Information booklet of the Ministry of Health Protection of the Russian Soviet Federal Socialistic Republic)
563. *Wacker, P. F.*: Quantifying hazardous microwave fields analysis. National Bureau od Standards. Techn. Note 391. Apr., 1970. See also [82], p. 197.
564. *Wacker, P. F., Bowman, R. R.*: Quantifying hazardous electromagnetic fields: scientific basis and practical considerations. IEEE Trans. Microwave Theory Techniques, vol. **MTT-19**, 178, 1971.
565. *Walawski, H.*: Prądy wielkiej częstotliwości (krótkie i ultrakrótkie fale) w biologii i medycynie. PZWL. Warszawa, 1952.
(High-frequency currents [short and ultrashort waves] in medicine and biology)
566. *Wąsowski, J.*: Anteny i fale. In [574], p. 37.
(Antennas and waves)
567. *Weiss, M. M., Mumford, W. W.*: Microwave radiation hazards. Health Phys. **5**, 160, 1961.
568. *Westgate, R.*: U. S. lights fire under microwave ovens. Electronics 123, 1969.

569. *Wever, R.*: Einfluss schwacher elektromagnetischer Felder auf die Circadiane Periodik des Menschen. Naturwissenschaften **55**, 29, 1968.
(Influence of weak electromagnetic fields on the circadian cycle of humans)
570. *Williams, D. B., Monahan, J. F., Nicholson, W. J., Aldrich, J. J.*: Biologic effects studies on microwave radiation. Time and power threshold for the production of lens opacities by 12.3 cm Microwaves. IRE. Trans. Med. Electron. New York, vol. **PGME-4**, 1956, p. 47.
571. *Williams, D. B., Monahan, J. P., Nicholson, W. J., Aldrich, J. J.*: Biological effects studies on microwave radiation. Time and power thresholds for the production of lens opacities by 12.3 cm microwaves. AMA Arch. Ophthalmol. **54**, 863, 1955.
572. *Witkowicz, J.*: Pomiary gęstości strumienia mocy mikrofalowej dla celów bhp. In [574], p. 96.
(Measurements of microwave power densities for hygiene and health protection purposes)
573. *Witkowicz, J.*: Efekty biologiczne i profilaktyka ekspozycji mikrofalowej. Prace Przemysłowego Instytutu Telekomunikacji **34**, 47, 1961.
(Biologic effects and prophylaxis of microwave exposure)
574. *Witkowicz, J.* (ed.), collective monograph: Higiena i bezpieczeństwo pracy w polu elektromagnetycznym mikrofal. Przemysłowy Instytut Telekomunikacji. Warsaw, 1964.
(Work hygiene and safety in electromagnetic microwave fields)
575. *Woods, D.*: Standard intensity electromagnetic field installation for calibration of radiation hazard monitors from 400 MHz to 40 GHz. Non-ionizing Radiation **1**, 9, 1969.
576. *Worden, R., Herrik, J., Wakim., Krusen, F.*: The heating effects of microwaves with and without ischaemia. Arch. Phys. Med. **29**, 751, 1948.
577. *Van Everdingen, W. A.*: Moleculare veranderingen en structurwizigungen teng wolge van bestraling met golven van Hertz met een goltlengte van 10 cm. Nederland Tijdschr, Geneeskunde **85**, 3094, 1941.
(Molecular structural changes induced by irradiation with hertzian waves of 10-cm wavelength)
578. *Van Everdingen, W. A.*: Moleculare veranderingen tengevolge van bestraling met golven van Hertz met een frequentie van 1875 megahertz. Nederland Tijdschr. Geneeskunde **84**, 4370, 1940.
(Molecular changes induced by irradiation with hertzian waves of 1875 MHz frequency)
579. *Van Everdingen, W. A.*: Veranderigen in de physisch-chemische constitute von organische vorbinden door stralenverking mede in varband met het kankenproblem. Nederland Tijdschr. Geneeskunde, **87**, 406, 1949.
(Changes in the physico-chemical constitution of organic compounds induced by radiation and the relationship to the cancer problem)
580. *Van Everdingen, W. A.*: Sur l'alteration moleculaire et structurale par irradiation avec des ondes hertziennes de 16 et 10 centimetres (1875 et 3000 MHz). 2. Transformations moleculaires. Rev. Belge. Sci. Med. **5**, 261, 1946.
(Molecular and structural alterations induced by irradiation with hertzian waves of 16 and 10 cm (1875 and 3 000 MHz) 2. Molecular transformations)
581. *Valtonen, E. J.*: Giant mast cells — a special degenerative form produced by microwave radiation. Exp. Cell Res. **43**, 221, 1966.
582. *Vassiliadis, A.*: Ocular damage from laser radiation. In: Wolbarsh, L. M. (ed.) Laser applications in medicine and biology, Vol. I, p. 125. Plenum Publishing Corporations, New York, 1971.
583. *Van Ummersen, C. V.*: The effects of 2450 Mc radiation on the development of the chick embryo. In [427], p. 201.
584. *Van Ummersen, C. V.*: An experimental study of developmental abnormalities induced in the chick embryo by exposure to radiofrequency waves. Ph. D. Dissertation. Dep. Biol. Tufts Univ., Medford, Mass. 1963.
585. *Van Ummersen, C. V., Cogan, F. C.*: Effects of microwave radiation on lens epithelial cells. In [82], p. 122.

586. *Vartanov, S. A.*: Vlijanie SVTch izlutchenii na immunobiologitcheskuju reaktivnost. Voj. Med. Zhurnal **11**, 52, 1969.
(The effect of exposure to ultra-high-frequency waves on immunobiologic reactivity)
587. *Vogelman, J. H.*: Physical characteristics of microwave and other radio frequency radiation. In [82], p. 7.
588. *Vogelhut, P. Q.*: Interaction of microwave and radiofrequency radiation with molecular systems In [82], p. 98.
589. *Volkova, A. P., Smurova, E. I.*: Vlijanie elektromagnitnych polei radiotchastot na fagocitoz i infekcjonnye vozpalenie u kryz. Gig. Santaria **9**, 107, 1969.
(Effects of electromagnetic radiofrequency fields on phagocytosis and infectious inflammation in rats)
590. *Voss, W. A. G.*: Microwave hazard control in design. In [82], p. 217.
591. Vremmenye sanitarnye pravila pri rabote s generatorami santimetrovych voln. Ministerstwo Zdravochranienia SSSR. 26.XI.1958 nr 273–58.
(Temporary sanitary rules for working with generators of centimeter waves) Ministry of Health Protection of the USSR. In [294], p. 121.
592. *Yagi, K.*: Local aplastic bone marrow induced by microwave irradiations, especially histological and histochemical studies. Nippon Acta Radiol. **30**, 184, 1970.
593. *Yao, K. T. S., Jiles, M. M.*: Effects of 2450 MHz microwave radiation on cultivated rat kangaroo cells. In [82], p. 123.
594. *Zaret, M. M., Cleary, S. F., Pasternack, B., Eisenbud, B., Schmidt, H.*: A study of lenticular imperfections in the eyes of microwave workers and a control population. Final Report RADC-TDR-63-125. N.Y. Univ. Medical Center, New York, 1963.
595. *Zaret, M. M., Kaplan, I. T., Kay, A. M.*: Clinical microwave cataracts. In [82], p. 82.
596. Zarządzenie Ministerstwa Zdrowia i Opieki Społecznej z dnia 20.VIII.1963. w sprawie warunków zdrowia wymaganych od pracowników narażonych na działanie pola elektromagnetycznego mikrofal. Monitor Polski Nr 66, poz. 328, 1963.
(Order of the Minister of Health and Social Welfare of Aug. 20, 1963 concerning health status requirements for workers exposed to microwave irradiation)
597. Zarządzenie Ministra Zdrowia i Opieki Społecznej z dnia 9.08.72 w sprawie określenia pól elektromagnetycznych w zakresie mikrofalowym oraz dopuszczalnego czasu pracy w strefie zagrożenia. Dziennik Urzędowy Ministerstwa Zdrowia i Opieki Społecznej Nr. 17, poz. 78. Sept. 20, 1972.
(Order of the Ministry of Health and Social Welfare, of Aug. 9, 1972, concerning the determination of electromagnetic fields and the permissible duration of work within the hazardous zone. Official Journal of the Ministry of Health and Social Welfare)
598. *Zakrzevskij, S. J., Kazelin, O. N.*: Metodika rostcheta zashtchitnych zon na poziciacj radiolokacyjnych stancji. Voj. Med. Zhurnal **12**, 42, 1966.
(Methods for calculation of safe zone in radar installations).
599. *Zimmer, R. P., Ecker, H. A., Popovic, V. P.*: Selective electromagnetic heating of tumors in animals in deep hypothermia. IEEE Trans. on Microwave Theory Techniques, vol. **MTT-19**, 238, 1971.
600. *Zyss, R., Boczyński, E.*: Zmiany morfologiczne i histochemiczne w wątrobie świnek morskich napromieniowanych mikrofalami. Biul. Wojsk. Akad. Med. **10**, 419, 1967.
(Morphologic and histochemical changes in the liver of guinea pigs irradiated with microwaves)
601. *Zyss, R., Boczyński, E., Sieliński, W.*: Wpływ mikrofal na morfologię niektórych narządów wewnętrznych świnki morskiej. Biul. Wojsk. Akad. Med. **10**, 429, 1967.
(The influence of microwaves on the morphology of certain internal organs of guinea pigs)
620. *Żydecki, S.*: Ocena przezroczystości soczewek oczu pracujących w zasięgu promieniowania.

mikrofalowego w porównaniu z grupami kontrolnymi. Thesis-Instytut Kształcenia Podyplomowego W.A.M. Warsaw, 1972.
(Evaluation of lens transparency in microwave workers as compared with control groups)
603. *Żydecki, S.*: Wyniki wstępnych badań i oceny przezroczystości soczewek u osób narażonych na działanie mikrofal. Klin. Oczna **38**, 31, 1968.
(Preliminary results and evaluation of lens transparency in individuals exposed to microwaves)
604. *Żydecki, S.*: Badania i ocena narządu wzroku u osób narażonych na działanie mikrofal (ze szczególnym uwzględnieniem zmętnień w soczewkach). Lek. Wojsk. **43**, 124, 1967.
(Examination and evaluation of the eye in individuals exposed to microwaves (with particular reference to lens opacities)

Added in proofs:

605. *Czerski P., Ostrowski K., Shore L.M., Silverman Ch., Suess M.J., Waldeskog B.* (eds): Biologic effects and health hazards of microwave radiation. Proceedings of an international symposium, Warsaw, 15–18 October 1973. Polish Medical Publishers. Warsaw 1974.
606. *Tyler P. E.* (ed): Biological effects of nonionizing radiation. Proceedings of a conference, New York 11–15 Febr 1974. Annals of the New York Acad. Sci.vol. 247, Febr 1975.
607. *East T. W. R., Ford J. D.*: Proceedings, Microwave Power Symposium. Waterloo, Canada May 28–30, 1975. International Microwave Power Institute. Edmonton 1975.
608. *Bawin S. M., Gavalas-Medici R. J., Adey W. R.*: Effects of modulated very high frequency fields on specific brain rhythms in cats. Brain Res. **58**, 365, 1973.
609. *Glaser Z. R.*: Bibliography of Reported Effects (phenomena) of Microwaves. Parts I–VII. NTIS Springfield, Va. 1973–1976.
610. *Adey W. R.*: Effects of Electromagnetic Radiation on the Nervous System. In: [606] p. 15.
611. *Bawin S. M., Kaczmarek L. K., Adey W. R.*: Effects of Modulated vhf Fields on the Central Nervous System. In: [606] p. 74.
612. "Effects of Weak Electromagnetic Fields on the Nervous System". Adey W. R. and Bawin S. M. (eds.). Neurosciences Research Bull. No XIX, Boston 1976.
613. *Grodsky I. T.*: Possible Physical Substrates for the Interaction of Electromagnetic Fields with Biologic Membranes. In: [606] p. 117.
614. *Czerski P., Paprocka-Slonka E, Stolarska A.*: Microwave Irradiation and the Circulation Rhythm of Bone Marrow Cell Mitoses. J. Microwave Power **23**, 32, 1974.

Subject Index to References

This index was appended with the aim of facilitating the selection of references for complementary or specialized reading. Monographs, symposia, and general surveys on microwave bioeffects and related subjects are listed under separate headings, specialized surveys and books under particular headings. The letters that follow the reference number designate:

E — English
/O — other language
P — Polish
R — Russian
S — survey

In certain instances translations are listed, the first letter indicates the language of the original, the second the language into which the publication was translated.

Complementary reading

Biological effects:
  comprehensive monographs: 1 R, 72 R, 185 R, 336/OE, 378 P, 425 RE, 438 RPE, 565 P, 574 P,
  magnetic fields: 33 E, 38 E
  biologic reference data: 54 E, 502 E, 505 E
Surveys: 11 E, 17/O, 20-21 P, 26 P, 51 E, 55-56/O, 80 E, 139/O, 204 E, 216 P, 217 E, 225-226 E, 228 E, 239 E, 253 E, 254/O, 258 E, 298 E, 351-352 E, 354 E, 357-358 E, 397 E, 416 E, 440 R, 442 R, 444 R, 450 E, 458 E, 462 E, 469 E, 474 E, 490 E, 510 R, 564 E
English surveys of USSR clinical findings: 127-130, 219
Symposia: 82 E, 139/O, 195 R, 294 RE, 418 E, 424 R, 427 E, 518 E, 605 E, 606 E, 612 E
Textbooks on microwave theory and techniques: 4 E, 97 E, 293 E, 408 ER, 476 ER
  on diathermy: 472/O, 477 E
  on radiobiology: 233 E
Physics: 287 (reference data)

Microwave Bioeffects and Related
Subjects: Specialized References

Absorption of microwave energy
  dielectric properties: 88 E, 89 E, 90 E, 137 E, 152 E, 245 E, 475 E, 482 ES, 483 ES
  distribution in bioobjects: 6 E, 8/O, 14 E, 15 E, 23/O, 25 P, 48/O, 49 E, 154 E, 209 E, 210 E, 211 E, 212 E, 230 E, 295 E, 299–303 E, 493 E, 563 E, 564 E
  mechnisms: 5 E, 6 E, 91 E, 252 E, 253 ES, 290 R, 443 ER, 449 E, 473 E, 474 E, 476 ES, 478–484 ES, 493 E, 563–564 E, 577–580/O, 582 E, 588 E
Adaptation, desadaptation: 208 R, 517 R
Analysis of the relationship of bioeffects
  field configuration: 3 E, 6 E, 14 E, 493 E, 564 E
  (see also absorption)
Auditory effects: 164 E, 166 E, 167 E, 168 E, 264 E
Behavioral effects: 163 E, 164 E, 165 E, 220 E, 256 ES, 264 R, 265 R, 269 E, 270 E, 280 ES, 281 E, 282 E, 291 R, 327 R, 536 E, 612 E
Biochemical findings and metabolism
  blood: 7 P, 12 E, 140 P, 192–193 R, 260 E, 291 R, 404–405 R, 453 E, 463/O, 498 R, 521–524 P,

529 RES, 530–533 R
man: 175 R, 179 RE
tissue: 7 P, 74 P, 79 P, 111 E, 123 P, 231/O, 248 E, 249/O, 262 R, 350 E, 389/O, 404 R, 585 E

Biological rhythm: 27 P, 369 P, 569/O, 585 E

Birds: 141 E, 276 E, 537 E

Blood and blood-forming system
  experimental: 20 P, 22 P, 27 P, 101 R, 102 E, 117 E, 119 E, 154 E, 188 R, 193 R, 194 R, 202 P, 203 P, 238 E, 244 R, 263 R, 352 E, 356 E, 357 E, 358 E, 360–363 E, 389/O, 436 E, 437 E, 495 P, 496 P, 504 E, 535 P, 592/O,
  man: 24 P, 34 E, 101 R, 102 E, 202 P, 203 P, 214 P, 215 P, 252a P, 308 E, 322 R, 387/O

Cardiovascular effects:
  experimental: 93 E, 94 E, 95 E, 96 E, 170 E, 183 E, 261 E, 311–313 R, 406 R, 414–415 E, 428–429 E, 445–446 R, 451 E, 456 E, 576 E
  man: 125 R, 177 E, 196 P, 511 RE

Chromosomes, cellular effects and mitosis: 28 P, 69 E, 74 P, 100 E, 101 R, 102 E, 221 E, 222 E, 223 E, 244 R, 248 E, 365 E, 369 P, 370 PS, 372 ES, 375 PS, 471 E, 506 P, 507 E, 544 E, 582 E, 585 E, 593 E
  (see also blood and blood-forming system, tissue culture)

Clinical syndromes of microwave exposure: 103 P, 135 R, 136 R, 148 E, 273 RS, 348a E, 367 P

Combined effects (microwaves plus X-rays and others): 238 E, 358 ES, 360 E, 361 E, 362 E, 423 RE, 549 E

Conditioned reflexes: 1 R, 57 R, 72 RS, 184 R, 189 R, 194 R, 198 R, 213 R, 247 R, 314–317 R, 319–320 R, 321 RE, 327 R, 382–383 R, 419–421 RE, 512-513 R, 525–527 R
  (see also nervous system)

Cumulative effects: 19–20 P, 22 P, 64 ES, 70 ES, 78 P, 101 R, 102 E, 111 E, 116–117 E, 142–143 PE, 350 E, 382 P, 495–496 P, 504 E,

Diathermy: 209–210 E, 212 E, 230 E, 295 E, 299–303 E

Digestive tract: 158–159 R, 198 R, 430 R, 509 RE, 548/O

Drugs and microwaves: 30 PE, 93–94 E, 96 E, 142 PE

Electrodes (see also metallic inclusions): 144 P, 164-165 E, 169 E

Endocrine system: 368 PS, 370 PS
  adrenals, experimental: 94 E, 192 R, 304 R, 305 P, 533 R
  man: 274 R
  hypophysis, experimental: 555 R
  pineal body: 76 P
  thyroid, experimental: 32 R
    man: 124 P, 126 R, 501 R
  varia: 140 P

Experimental methods, see also measurements: 171 E, 347 E, 439 R, 519 E

Eye:
  clinical cataracts: 81 E, 229 E, 292a E, 373 E, 433 PS, 534 R, 595 ES
  experimental: 15–16 E, 45 R, 46–47 E, 64 ES, 65–70 E, 107–108 P, 109–11 E, 138 E, 373 E, 413 P, 454 E, 457 E, 570–571 E
  lens opacities man: 45 R, 81 E, 250 P, 285 P, 325 P, 373 E, 459 ES, 545 E (diathermy hazards), 594 E, 595 ES, 602–604 P
  radar operators: 41/O, 61/O, 178 E, 268 E, 385–386 P, 459 ES
  retinal changes, man: 255 R, 338 E, 547 E

Genetic effects, fertility, fetal development, reproductive functions:
  experimental: 42 R, 69 E, 193–194 R, 276 E, 318 R, 365 E, 375 PS, 381 P, 388/O, 391 E, 414–415

E, 435 R, 436 E, 486 E, 504 E, 577–578/O, 584 E
man: 83/O, 155 E, 375 PS, 497 E
Genital organs, male and female: 74 P, 77–78 P, 132 P, 147 E, 202–203 P, 205–206 E, 231/O, 243 E, 323–324 P
Hazards: 17/O, 51 E, 53 E, 55–56/O (see also prevention, safe exposure limits)
Health status of persons occupationally exposed: 34 E, 44/O, 104 PS, 106 E, 133/O, 135–136 R, 140 P, 143 RE, 176 R, 181 R, 184 RS, 186 E, 202–203 P, 207 R, 235–236 P, 252a P, 270a–270b/O, 273 RS, 275 P, 283 P, 288 P, 308 E, 322 R, 329 P, 335 P, 338 E, 339 P, 370 PS, 374, P, 377 P, 387/O, 409a P, 547 E, 552 R, 562 RS 596 P, (see also entry particular organs, systems, and tissues subentries under man)
Irradiation CW, PW, intermittent: 59 R, 114 E, 116 E, 122 E, 495–496 P
Measurements: 49–50 E, 98 E, 253 ES, 395 E, 492 R, 519 E, 563–564 E, 572 P, 575 E, 598 R (see prevention, protective equipment, and measures, hygienic analysis of working conditions)
Metabolism; see biochemical findings
Metallic inclusions: 14 E, 396/O
Microwave ovens: 266–267 P, 422 E, 494 E, 503 E, 550 E, 568 E
Mitosis; see chromosomes
Morphologic findings (histology), gross findings: 1 R, 19–20 E, 29 P, 31–32 PE, 74 P, 157 E, 191 R, 193 R, 197 E, 205–206 E, 234/O, 249/O, 309 E, 326/O, 379 P, 403 P, 409 E, 431 R, 554–557 R, 592/O, 600–601 P
(see also entries under particular organs or systems, subentry experimental)
Muscles: 29 P, 295 E
Nervous system
central experimental: 8/O, 9–10 E, 12 E, 18 P, 19 P, 20 P, 30–31 PE, 37 R, 57 R, 59–60 R, 71 R, 72 RS, 73 R, 84–86 E, 142 PE, 144 P, 162–163 E, 189–191 R, 193–194 R, 213 R, 218/O, 247 R, 271 E, 314–317 R, 318–320 R, 321 RE, 342 E, 344 E, 409 E, 485–486 E, 493 E, 554–556 R
man: 58 R, 135–136 R, 143 PE, 176 R, 181 R, 270a–270b/O, 335 P, 339 P, 432 R, 608 E, 611–613 E
peripheral, experimental: 43 R, 160 E, 174 R, 259 R, 307 R, 334/O, 340–347 E (see also conditional reflexes behavioral effects)
Non-thermal effects
12 ES, 13 E, 19 P, 22 P, 30–31 PE, 43 R, 57–58 R, 60 R, 62–64 E, 69 E, 70 ES, 71 R, 73 R, 78 P, 100 E, 101 R, 102 E, 111 E, 117 E, 143 PE, 162 E, 164–165 E, 170 E, 181 R, 189–191 R, 213 R, 260 E, 350 E, 360–361 E, 363 E, 365 E, 391 E, 445–446 R, 454–455 E, 457 E, 471 E, 486 E, 493 E, 519 E, 577–580/O, 585 E, 613 E
Pearl chain formation: 200–201 E, 224 E, 544 E
Prevention, protective equipment, and measures, hygienic analysis of working conditions: 40 R, 51 E, 53 E, 55–56/O, 61/O, 98 E, 99 R, 104–105 P, 125 R, 145 E, 156 RE, 172–173 R, 178–179 E, 180 R, 182 RE, 184 RS, 186 E, 204 ES, 241 E, 268 E, 272 RE, 288–289 P, 292 R, 310 P, 328–331 P, 348 ES, 371 ES, 376 PS, 395 E, 397 E, 401–402 E, 412 R, 417 P, 434 E, 441 RS, 459 ES, 465–467 E, 490 ES, 492 R, 500/OS, 543 E, 545–546 E (reference data) 551 R, 560 R, 568 E, 573 P, 575 E, 590 E, 596 P, 598 R
Safe exposure limits, radiation protection guides: 35 E, 87 E, 104 P, 146 E, 289 P, 332–333 E, 337–338 E, 349/O, 351 E, 354–355 E, 366 E, 380 E, 384 P, 398–400 E, 416 ES, 422 E, 426 RE, 447 E, 460–461 P, 465–468 E, 476 ES, 490–491 ES, 520 ES, 528–539 ES, 543 E, 559 E, 561 E, 567 E, 591 R, 597 P, R, 598 R
Thermal effects: 2–3 E, 5–6 E, 8/O, 14–15 E, 20 P, 23/O, 25 P, 36 E, 48/O, 52/E, 91–92 E, 95 E, 112 E, 113–117 E, 119 E, 121 E, 123 P, 147 E, 149–151 E, 154 E, 157 E, 160–161 E, 187 RE, 188–189 R, 209–212 E, 228–230 E, 232 ES, 234/O, 237 E, 240–241 E, 246 E, 257/O, 258 ES, 286 E, 299–303 E, 340–347 E, 353 E, 356 E, 357 ES, 359 E, 361–364 E, 393–394 R, 407/O, 410–411 E, 448 E, 453 E, 487–488 E, 489 R, 493 E, 499–500 R, 514 RE, 553 R, 599 E

influence of environment: 118 E, 122 E, 228 E, 232 E, 256 E, 257/O, 258 ES, 264 E, 397 ES, 398–400 E, 470 E
Thermal versus nonthermal effects: 15–16 E, 20 P, 26 PS, 390–392 E, 450 RS, 558 R
Thermography: 209–212 E, 230 E
Tissue culture: 62–63 E, 100 E, 296–297 E, 414–415 E, 471 E, 506 P, 507 E, 584 E, 593 E
Varia: 39 E, 134 E, 197 E, 199 E, 277–279 P, 372 E, 528 R, 542 R, 586 R, 589 R, 599 E

Subject index

Absorbed power density, 47, 51, 55, 57, 61, 62, 86, 88
Absorption relative cross section (see: Cross section)
Acetylcholinesterase, 108—109, 113
Adrenals, 123, 159
Amylase, 125
Analeptic effects, 95
Antenna, 34—40
Attenuation constant, 47
Average power, 41

Behavioral effects, 100—102
Biological membranes, 68—70
Birds, 116, 130
Blood pressure, 118—121, 158
Blood proteins, 125
Brain lesions, 97—98, 105—106, 113

Caffeine, 93
Carbohydrate metabolism, 122
Cardiazole, 163
Carcinotron, 19
Cardiasol, 105
Cataractogene dose, 147, 148
Cataract, 165, 171, 173
Cells in vitro, 132
Centimeter waves, 13
Central nervous system, 57, 58, 68, 69, 91, 92, 114, 116, 122
Chlorpromazine, 105
Cholinesterase, 93 (see also: Acetylcholinesterase)
Chromosomal aberrations, 132, 141
Circadian rhythms, 133, 140, 142
Circulatory effects, 117, 122, 155
Complains subjective, 159, 162
Conditioned reflexes, 96—100, 115, 119, 144, 145
Conduction currents 64—65
Continuous wave, 17, 84, 92, 108, 110, 149, 178
Cooling by blood flow, 90
Cross section, 55, 58, 170
Cumulative effects, 148
Curare 107
Cytochrome oxidase, 110, 126

Decimeter waves, 13
Depth of penetration (see: Skin depth)
Desadaptation, 118
Dielectric constant, 27, 50, 64
Diffraction, 12

Direct effects, 72
Dispersion, 12
Duty cycle, 41

ECG, 118, 121, 156, 157, 164
EEG, 92, 94, 103—105, 107, 108, 156, 157, 163
Effective conductivity, 65
Effective irradiation time, 42, 84, 179
Electrodes, 62, 94, 102
Electric fields, 24, 30
Electromagnetic fields, 12, 24, 31
Electromagnetic field theory, 12
Electromagnetic radiation spectrum, definition 11, 12, 13
Electromagnetic wave, 30, 31
Energy density (see: Power density)

Far field, 31, 37, 38, 59
Fetal development, 130, 131, 134
Frequency, 24, 28

Gain, 34
Ganglioplegic drugs, 93, 118
Glutathione, 125

Hazards, 7, 11, 14, 15, 16, 20, 21, 43, 59, 170, 186
Heat conduction, 90
High-energy radiation (see: Ionizing radiation)
Hypophysis, 122
Hypothalamus, 122, 123

Impedance, 30
Inborn defects in man, 134—135
Infrared (IR), 11
Intended radiation, 20, 21, 34
Intermittent irradiation, 85
Internal field distribution, 74, 90, 135
(See also: Planar model; Spherical model; Termography)
Ionizing radiation, 11, 17, 74, 76, 138
Irradiation cycle rate, 42, 84

Klystron, 17

Leakage, 21, 38
Lens opacities, 159, 165, 166, 168, 171, 177
Lipase, 125
Loss tangent, 65
Luminal 105

Lymphoblastoid transformation, 133, 139

Magnetic field intensity, 24, 30
Magnetron, 15, 16
Mating behavior, 128, 129
Meter waves, 13
Microwave, bands, 14
  generation, 14, 20, 26, 40
  heating, 15, 65, 73
  region, 11
  sources, 43
  uses, 16, 44
Millimeter waves, 13, 14
Mitotic abnormalities, 132, 133, 140, 141
Modulation, 11, 18

Near field, 31, 38, 59
Nonintended radiation, 20, 21, 34, 154
Non-stationary fields, 42
Nonthermal effects, 42

Oxygen consumption, 124

Peak power, 17, 47
Permeability, 27
Pentatetrasol, 107 (see also: Cardiasol)
Period, 27
Permittivity (see Dielectric constant, 27)
Phase shift, 44
Photon, 32, 33
Planar model, 48—49, 53, 54, 63
Primary effects (see: Primary interaction)
Primary interaction, 46, 63, 70, 77
Power density, 22, 30, 38, 39
Pulsed waves, 17, 40, 41, 84, 92, 108, 110, 119, 138, 144, 145, 149, 178
Pupal development, 130, 131
Pyridoxine, 118

Quantum effects, 12, 66, 70, 71

Reflection, 12
Reflection coefficient, 48, 50
Relaxation time, 64
Reserpine, 118
Risks (see: Hazards)

Safe exposure limits, 7
Secondary effects, 72, 77
Skin depth, 48, 50

Specific heating, 65—66, 73
Spherical model, 55—57, 63
Standing waves, 49, 51
Stationary fields, 42
Stomach ulcers, 136
Stress, 46, 74
Strychnine, 105
Succinic acid dehydrogenase, 110
Succinic dehydrogenase, 126

Temperature gradients, 89
Temperature humidity index, 174, 175
Testes 124—128
Thermal effects, 65, 72, 73, 78, 86, 87, 124
Thermography, 60—63
Thyroid, 124—126
Transmission line, 35

Traveling-wave tubes, 19
Tubocurarine, 105

Ultraviolet (UV), 11

Volume heating, 66

Waveguide, 35, 36
Wavelength, 11, 14, 27, 28, 47, 50

X-rays (see: Lonizing radiation)

QP
82.2
M5
B37

SEP 27 1977